Turbulence
A Tentative Dictionary

AF393369

NATO ASI Series

Advanced Science Institutes Series

A series presenting the results of activities sponsored by the NATO Science Committee, which aims at the dissemination of advanced scientific and technological knowledge, with a view to strengthening links between scientific communities.

The series is published by an international board of publishers in conjunction with the NATO Scientific Affairs Division

A	Life Sciences	Plenum Publishing Corporation
B	Physics	New York and London
C	Mathematical	Kluwer Academic Publishers
	and Physical Sciences	Dordrecht, Boston, and London
D	Behavioral and Social Sciences	
E	Applied Sciences	
F	Computer and Systems Sciences	Springer-Verlag
G	Ecological Sciences	Berlin, Heidelberg, New York, London,
H	Cell Biology	Paris, Tokyo, Hong Kong, and Barcelona
I	Global Environmental Change	

Recent Volumes in this Series

Volume 334—Frontier Topics in Nuclear Physics
edited by Werner Scheid and Aurel Sandulescu

Volume 335—Hot and Dense Nuclear Matter
edited by Walter Greiner, Horst Stöcker, and André Gallmann

Volume 336—From Newton to Chaos: Modern Techniques for Understanding
and Coping with Chaos in N-Body Systems
edited by A. E. Roy and B. A. Stevens

Volume 337—Density Functional Theory
edited by Eberhard K. U. Gross and Reiner M. Dreizler

Volume 338—Electroweak Physics and the Early Universe
edited by Jorge C. Romão and Filipe Freire

Volume 339—Nonlinear Spectroscopy of Solids: Advances and Applications
edited by Baldassare Di Bartolo

Volume 340—Confined Electrons and Photons: New Physics and Applications
edited by Elias Burstein and Claude Weisbuch

Volume 341—Turbulence: A Tentative Dictionary
edited by P. Tabeling and O. Cardoso

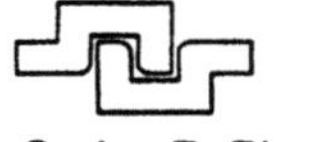

Series B: Physics

SPECIAL PROGRAM ON CHAOS, ORDER, AND PATTERNS

This book contains the proceedings of a NATO Advanced Research Workshop held within the program of activities of the NATO Special Program on Chaos, Order, and Patterns.

Turbulence

A Tentative Dictionary

Edited by

P. Tabeling and O. Cardoso

Laboratoire de Physique Statistique
Ecole Normale Supérieure
Paris, France

Springer Science+Business Media, LLC

Proceedings of a NATO Advanced Study Institute on
Turbulence, Weak and Strong,
held August 2–14, 1993,
in Cargèse, France

NATO-PCO-DATA BASE

The electronic index to the NATO ASI Series provides full bibliographical references (with keywords and/or abstracts) to more than 30,000 contributions from international scientists published in all sections of the NATO ASI Series. Access to the NATO-PCO-DATA BASE is possible in two ways:

—via online FILE 128 (NATO-PCO-DATA BASE) hosted by ESRIN, Via Galileo Galilei, I-00044 Frascati, Italy

—via CD-ROM "NATO Science and Technology Disk" with user-friendly retrieval software in English, French, and German (©WTV GmbH and DATAWARE Technologies, Inc. 1989). The CD-ROM also contains the AGARD Aerospace Database.

The CD-ROM can be ordered through any member of the Board of Publishers or through NATO-PCO, Overijse, Belgium.

Library of Congress Cataloging-in-Publication Data

NATO Advanced Study Institute on Turbulence, Weak and Strong (1993 :
  Cargèse, France)
    Turbulence : a tentative dictionary / edited by P. Tabeling and O.
  Cardoso.
       p.   cm. -- (NATO ASI series. Series B, Physics ; v. 341)
     "Published in cooperation with NATO Scientific Affairs Division."
     Includes bibliographical references and index.
     ISBN 978-1-4613-6106-0   ISBN 978-1-4615-2586-8 (eBook)
     DOI 10.1007/978-1-4615-2586-8
     1. Turbulence.   I. Tabeling, P.   II. Cardoso, O.   III. North
  Atlantic Treaty Organization.   Scientific Affairs Division.
  IV. Title. V. Series.
  TA357.5.T87N385   1995
  620.1'064--dc20   94-49720
                                                         CIP

Additional material to this book can be downloaded from http://extra.springer.com.

10 9 8 7 6 5 4 3

ISBN 978-1-4613-6106-0

© 1994 Springer Science+Business Media New York
Originally published by Plenum Press, New York in 1994

All rights reserved

No part of this book may be reproduced, stored in a retrieval system, or transmitted in any form or by any means, electronic, mechanical, photocopying, microfilming, recording, or otherwise, without written permission from the Publisher

SPECIAL PROGRAM ON CHAOS, ORDER, AND PATTERNS

SPECIAL PROGRAM ON CHAOS, ORDER, AND PATTERNS

Volume 341 — TURBULENCE: A Tentative Dictionary
edited by P. Tabeling and O. Cardoso

PREFACE

The present volume comprises the contributions of some of the participants of the
NATO Advance Studies Institute "Turbulence, Weak and Strong", held in Cargèse, in
August 1994. More than 70 scientists, from seniors to young students, have joined together to discuss and review new (and not so new) ideas and developments in the study of
turbulence. One of the objectives of the School was to incorporate, in the same meeting,
two aspects of turbulence, which are obviously linked, and which are often treated sep-
arately: fully developed turbulence (in two and three dimensions) and weak turbulence
(essentially one and two-dimensional systems). The idea of preparing a dictionary rather
than ordinary proceedings started from the feeling that the terminology of turbulence
includes many long, technical, poorly evocative words, which are usually not understood
by people exterior to the field, and which might be worth explaining. Students who start
working in the field of turbulence face a sort of curious situation: on one side, they are
aware that turbulence is related to the disordered, churning flows of torrents, the powerful movements of water in the oceans, the violent jet streams in the troposphere, the
solar eruptions, and they are certainly excited to pierce the mystery of this fascinating,
omnipresent phenomenon. On the other side, in many cases, they will use most of their
energy investigating quantities as sophisticated as high order structure functions of the velocity increments, spectra of the enstrophy, or the phase of some slowly varying complex
variable. It is certainly necessary to consider these quantities for studying turbulence.
High order structure functions underline the contribution of rare events, enstrophy decay
is a measure of the filamentation process in 2D turbulence, and most of the weakly turbulent flow pattern can be described in term of slowly varying complex variables. Also, there
are simpler words, such as vortex filaments, or vortex ribbons, which are currently used
in the field and whose physical meaning is more straightforward to capture. But clearly,
we had the feeling that it could be useful to change a little the format of the proceedings,
and try to build, tentatively, a dictionary, where several concepts are explained, so as to
help young researchers and non experts being introduced in the field. This is the goal of
this volume.

The organization of this book mimics that of an encyclopedia, and each article is devoted
to reviewing a concept in a few pages. In most cases, the articles are accessible to readers
unfamiliar with the concepts of turbulence. There are 23 entries, covering a significant
part of the important concepts used in the domain; obviously, several words, currently
used in the field, are missing, but the scope covered by the present book seems wide
enough to justify the title "Tentative Dictionary on Turbulence".

We thank the financial support by NATO, CNRS, DRET, Ministère des Affaires Etran-
gres, LVMH, CRAY, CNES. The ASI was organized by one of us (P.T.) and V. Emsellem
who has also contributed to the preparation of these proceedings. We thank C. Philippe
and A. Manchon for their help at various stages of the organization of the School, and
D. Marteau, O. Paireau, H. Willaime and F. Belin for their careful reading of the manu-

scripts; we are also pleased to acknowledge M.F. Hanseler and the staff members of Institut d'Etudes Scientifiques de Cargèse for their considerable contribution to the success of the School.

O. Cardoso, P. Tabeling

CONTENTS

Decaying Two-dimensional Turbulence

W.R. Young

Scripps Institution of Oceanography,
La Jolla CA 92093-0230, USA

INTRODUCTION

Decaying two-dimensional turbulence refers to the solution of the two-dimensional, incompressible Navier-Stokes equation as an unforced initial value problem. This is an instructive counterpoint to the complementary problem of forced two-dimensional turbulence. In both cases one is concerned with high Reynolds number flows in which inertia dominates the other processes (viscosity and forcing) over many decades of length scales.

In terms of the streamfunction $\psi(x, y, t)$ the velocity (u, v) and the vorticity $\zeta = v_x - u_y$ are given by

$$(u, v) = (-\psi_y, \psi_x), \qquad \zeta = \nabla^2 \psi. \quad (1)$$

The sole governing equation is

$$\zeta_t + J(\psi, \zeta) = -\nu_0 \zeta + \nu_2 \nabla^2 \zeta, \quad (2)$$

where $J(\psi, \zeta) = \psi_x \zeta_y - \psi_y \zeta_x$ is the nonlinear advection of vorticity. If the terms on the right hand side of (2) are to be dissipative then $\nu_n \geq 0$.

In numerical solutions the domain is often a square of side $2\pi L$ and the boundary condition is that ψ is periodic in both x and y. For meteorologists and astrophysicists, the alternative of spherical geometry has many attractions. The initial condition might be constructed by using a random number generator to select the amplitude of each Fourier component of ψ. The variance at each wavenumber can then be adjusted to give any desired spectral shape to the initial conditions. The initial conditions do not have to be random: deterministic initial conditions, such as an unstable shear layer, produce nontrivial dynamical evolution. In the following I focus on the case of doubly–periodic geometry and random initial conditions.

Numerical solutions of the initial value problem outlined above show that well separated, almost axisymmetric, persistent vortices form from random initial conditions[1, 2, 3, 4]. Between the vortices there is a sea of small–scale, incoherent vorticity. After the emergence of the vortices the dynamics consists of two processes: *(i)* mutual advection of the well separated vortices in which Hamiltonian point-vortex dynamics is a good approximation[5] *(ii)* merger of like-signed vortices during close encounters[6].

The mergers occur when the irregular Hamiltonian motion jostles a pair of like signed vortices into proximity. The merger process occurs in about one vortex turnover time and it results in the expulsion of some vorticity from the vortex cores into the sea of small scale incoherent vorticity. Over very long times the irregular random motion of the vortices will affect all possible mergers so only two opposite signed vortices (a dipole) remain in the domain[7, 8].

The remainder of this review addresses two questions suggested by the above scenario: *(i)* What are the statistical properties of the vortex gas during the prolonged evolution towards the final dipole? *(ii)* What can be said (or conjectured) about the structure of the final dipole? This bipartite focus ignores at least one remarkable aspect of decaying two-dimensional turbulence which is the rapid emergence of axisymmetric vortices

Turbulence: A Tentative Dictionary
Edited by P. Tabeling and O. Cardoso, Plenum Press, New York, 1995

from random initial conditions. There has been no research on this fundamental process other than observations of it in many numerical solutions.

DISSIPATION AND THE INVARIANTS OF TWO DIMENSIONAL VORTICITY DYNAMICS

Theoretical arguments concerning decaying two-dimensional turbulence hinge on the inviscid invariants of (2). Now the left hand side of (2) is the material advection of vorticity on fluid particles. If one multiplies (2) by $F'(\zeta)$, where $F(\zeta)$ is any function of ζ, and then integrates over the domain then the result is

$$\frac{d}{dt} \int F(\zeta)\, dA = \int F'(\zeta) \mathcal{D}\, dA \qquad (3)$$

where $\mathcal{D}$ represents the dissipative terms on the right hand side of (2). If $\mathcal{D} = 0$ then (3) shows that there are an infinite number of vorticity invariants. In physical terms this infinitude of inviscid invariants results from the rearrangement of the vorticity distribution by material advection.

The dissipative terms on the right hand side of (2) deserve some comment before we proceed. Navier-Stokes dynamics has $\nu_0 = 0$. Many numerical solutions take $\nu_0 = \nu_2 = 0$ but use a "hyperviscosity" such as $-\nu_4 \nabla^4 \zeta$. This computational artifice ensures that the effects of damping are confined to the very highest wavenumbers in the calculation.

The possibility of experimental realization of decaying two-dimensional turbulence is discussed elsewhere in this volume. In the experimental context the term $-\nu_0 \zeta$ on the right hand side of (1) (sometimes called "bottom" or "Eckman" drag) is often as important as the lateral friction term $\nu_2 \nabla^2 \zeta$.

Finally, in certain analytic theories of two-dimensional turbulence, one considers the possibility that $\nu_0 = \nu_2 = 0$ so that the Reynolds number is infinite rather than merely very large[9, 10]. The adjective "decaying" may seem inappropriate in this idealized case. But if one admits the possibility of indefinitely small length scales

in the dynamical fields then the coarse grained versions of the vorticity invariants in (3) will be "dissipated" as the vorticity field is teased out into fine-scale filaments by material advection. Of course, evolutionary numerical solutions of (2) have finite spatial resolution so that this infinite Reynolds number limit is not computationally accessible. It is a plausible, but unproven, assumption that the large scale features of high Reynolds number solutions are similar to the structures which would be seen in solutions with infinite Reynolds number.

We now return to the discussion of the invariants of the vorticity equation. In addition to the vorticity invariants there is the energy

$$E(t) \equiv \int \frac{1}{2} \nabla \psi \cdot \nabla \psi\, dA. \qquad (4)$$

If (2) is multiplied by ψ and integrated over the domain then one finds

$$\frac{dE}{dt} = -2\nu_0 E - 2\nu_2 Z, \qquad (5)$$

where $Z(t)$ is the mean square vorticity or "enstrophy" corresponding to the choice $F(\zeta) = \zeta^2/2$ in (3). The crucial difference between two and three-dimensional turbulence is that in two dimensions the enstrophy is bounded by its initial value —there is no vortex stretching to create vorticity. If (2) is multiplied by ζ and integrated over the domain one finds that

$$\frac{dZ}{dt} = -2\nu_0 Z - \nu_2 \int \nabla \zeta \cdot \nabla \zeta\, dA < 0. \qquad (6)$$

Because $Z(t) < Z(0)$ the rate of energy dissipation on the right hand side of (5) must go to zero if the frictional parameters ν_0 and ν_2 are reduced. This is in stark contrast to the three-dimensional case. One says that energy is a "robust" or "rugged" invariant meaning that it is dissipated only on the very long frictional time scales.

THE VORTEX GAS

The evolution of the vortex gas towards the final dipole state suggests questions such as how does the number of vortices per unit area decay with time? How does the average radius and circulation of the vortex population evolve with time? How does the fractional area covered by vortices change with time? An automated "vortex census" of a numerical simulation has provided empirical information about these issues[11]. For instance, the number of vortices per area $\rho(t)$ decreases like $\rho(t) \sim t^{-\xi}$ where the exponent $\xi \approx 0.75$.

The simplest theoretical argument assumes that the energy per unit area $\mathcal{E} \equiv E/4\pi^2 L^2$ is the only robust invariant. If this is true then $\mathcal{E}$ is the only property of the initial conditions which can determine the statistical properties of the vortex gas. On dimensional grounds then $\rho(t) \sim 1/\mathcal{E}t^2$. This prediction is inconsistent with the results of numerical simulation. We conclude that in addition to E there must be other robust invariants that are inherited from the initial conditions.

Perhaps the simplest scaling assumption which is consistent with a detailed examination of the vortex merger process is that the extrema of the initial vorticity distribution are rugged invariants. It is the maxima and minima of the initial vorticity field which become the cores of the emergent vortex gas. And when two vortices merge the extreme values of the vorticity field survive the collision while the lower lying values of ζ are jettisoned into the sea of incoherent vorticity filaments.

If we denote the typical value of vorticity extrema by ζ_{ext} then there are now two rugged invariants: ζ_{ext} and $\mathcal{E}$. Since ζ_{ext}^{-1} has the dimensions of time one can now form both a length and a time scale from properties of the initial conditions: consequently dimensional arguments no longer determine $\rho(t)$. But if one assumes that $\rho(t) \sim t^{-\xi}$ then scaling arguments can be used to relate properties of the vortex population, such as the average radius, to the undetermined exponent ξ. For instance, this scaling theory predicts that the average vortex radius increases like $t^{\xi/4}$ and that the average vortex circulation increases like $t^{\xi/2}$. Comparison with numerical simulations supports these predictions. As yet there is no wholly convincing explanation for the value of the exponent $\xi \approx 0.75$.

THE FINAL DIPOLE

After all possible mergers have occurred there is complete segregation of the vorticity field into one positive vortex and one negative vortex. While the scaling theory outlined above does provide some information about the rate of approach to this final equilibrium[13] it provides little information about the structure of the final dipole. The simplest suggestion for determining the structure of the ultimate dipole is the "selective decay" hypothesis[14, 15] which says that the evolution minimizes the enstrophy $Z(t)$ as much as is consistent with a specified initial energy E. The solution of this simple variational problem is the gravest Helmholtz eigenmode (i.e. ζ is a constant multiple of ψ) of the domain. This prediction of the selective decay hypothesis is not consistent with numerical solutions: the vorticity amplitude and the enstrophy are larger, and the vortex radius smaller, than those of the Helmholtz eigenmode. All of these discrepancies are consistent with the assumptions of the scaling theory which takes account of ζ_{ext} as a second robust invariant.

Other statistical approaches to calculating the structure of the final dipole include a self-consistent mean field theory of point vortices[16] and, more recently, a statistical field theory for continuous vorticity distributions[9, 10]. The mean field theory of point vortices[16] predicts that the final dipole satisfies the "sinh-Poisson" equation:

$$c\zeta = sinh(|\beta|\psi). \qquad (7)$$

and this prediction is more consistent with the computational results than the predictions of selective decay hypothesis.

The statistical field theories for continuous vorticity distributions[9, 10] lead to more complicated functional relations between ζ and ψ than (7). Because these theories account for the infinitude of conserved vorticity invariants this complexity is unavoidable. Thus, although the point vortex mean field theory can be criticized because it places no constraints on the separation of opposite signed point vortices, at present (7) is the only simple and viable prediction for the structure that ultimately emerges from a random initial vorticity distribution.

Bibliography

[1] J.C. McWilliams, *J. Fluid Mech.* **146**, 21 (1984).

[2] A. Babiano, C. Basdevant, B. Legras and R. Sadourny, *J. Fluid Mech.* **183**, 379 (1987).

[3] P. Santangelo, R. Benzi and B. Legras, *Phys. Fluids A* **1**, 1027 (1989).

[4] M. Brachet, M. Meneguzzi, H. Politano and P. Sulem, *J. Fluid Mech.* **194**, 333 (1988).

[5] R. Benzi, S. Patarnello and P. Santangelo, *Europhys. Lett.* **3**, 811 (1987).

[6] M.V. Melander, N.J. Zabusky and J.C. McWilliams, *J. Fluid Mech.* **195**, 303 (1988).

[7] W.H. Matthaeus, W.T. Stribling, D. Martínez, S. Oughton and D. Montgomery, *Phys. Rev. Lett.* **66**, 2731 (1991).

[8] W.H. Matthaeus, W.T. Stribling, D. Martínez, S. Oughton and D. Montgomery, *Physica D* **51**, 531 (1991).

[9] R. Robert and J. Sommeria, *J. Fluid Mech.* **229**, 291 (1991).

[10] J. Miller, *Phys. Rev. Lett.* **65**, 2137 (1990).

[11] J.C. McWilliams, *J. Fluid Mech.* **219**, 361 (1990).

[12] G.F. Carnevale, J.C. McWilliams, Y. Pomeau, J.B. Weiss and W.R. Young, *Phys. Rev. Lett.*, **66**, 2735 (1991).

[13] G.F. Carnevale, J.C. McWilliams, Y. Pomeau, J.B. Weiss and W.R. Young, *Phys. Fluids A*, **4**, 1314 (1992).

[14] F.P. Bretherton and D.B. Haidvogel, *J. Fluid Mech.* **78**, 129 (1976).

[15] D. Montgomery, L. Turner and G. Vahala, *Phys. Fluids*, **21**, 757 (1978).

[16] D. Montgomery, W.H. Matthaeus, W.T. Stribling, D. Martínez and S. Oughton, *Phys. Fluids A* **4**, 3 (1991).

Experiments in 1D Turbulence

F. Daviaud
Service de Physique de l'Etat Condensé, CEA, Centre d'Etudes de Saclay
F-91191 Gif-sur-Yvette cedex, France

1D TURBULENCE does not exist in a strict sense. This expression is usually used for spatiotemporal chaos in systems in which the dynamical regimes depend on time and on one space variable only. In the following, we first define what we mean by (quasi) one-dimensional systems and give some general experimental considerations. The transition to spatiotemporal chaos is then analyzed in a model hydrodynamical system (Rayleigh-Bénard convection) which reveals a transition via spatiotemporal intermittency. Finally, other experiments in which 1D turbulence is observed are presented.

QUASI-ONE-DIMENSIONAL SYSTEMS

The study of quasi-one-dimensional -1D- hydrodynamical systems has recently received much interest. In these systems, a steady or traveling cellular (periodic in space) state, which depends only on one space variable, may destabilize and become chaotic when a control parameter is varied. The spatio-temporal chaos observed thus stands between the deterministic chaos of confined systems and the developed turbulence of extended ones.

Many experimental studies have been performed in 1D periodic systems. All these different systems are studied in geometries with one dimension larger than the others but they are not strictly speaking one-dimensional because their physical properties (velocity, temperature, ...) generally depend on the three dimensions of space. However, the 1D aspect can be obtained by different means: geometries with a small transversal aspect ratio, experiments dealing with an interface or consisting of a linear arrays of coupled units. For a given value of a control parameter, these systems show a pattern (stationary or propagating) which is periodic in one direction of space (x). When the control parameter is varied, the evolution of this pattern does not depend, in first approximation, on the other directions of space but only on x and on time.

This 1D aspect can be very well illustrated in the case of Rayleigh-Bénard convection. Figure 1 shows a picture of a fourfold cell with different transverse aspect ratios $\Gamma_y = L_y/d$ (where d is the depth of the layer) but for the same value of the control parameter (here the Rayleigh number). On this figure, one can see, and it has been demonstrated quantitatively, that for $\Gamma_y < 0.5$ the dynamical regimes only depend on x, whereas they depend on x and y for $\Gamma_y > 0.5$.

The experiments can now be classified according to their 1D aspect:

- *narrow gap geometries:* Rayleigh-Bénard convection[1, 2], convection in binary mixtures[3], parametric instability[4].

- *interfaces:* directional viscous fingering[5], gravitational instability of a 1D liquid film[6], directional solidification of a moving nematic-isotropic interface[7], thin-film lamellar eutectic growth[8], Taylor-Dean system[9], fluid layers with a free surface locally heated[10] or submitted to a horizontal temperature gradient[11].

Turbulence: A Tentative Dictionary
Edited by P. Tabeling and O. Cardoso, Plenum Press, New York, 1995

- arrays of coupled units: line of vortices[12], forced convection[13, 14], array of von Kármán wakes[15].

Much theoretical and numerical work has also been devoted to the study of 1D spatiotemporal chaos in 1D amplitude equations, coupled map arrays and 1D cellular automata[16, 17]. But the situation is far less clear than in the case of (temporal) chaos, where well-defined routes to chaos have been predicted and observed, and no general theory is today available for extended systems. Two types of transition to turbulence can be distinguished. The supercritical transitions are those where disorder appears progressively, as a small perturbation of the regular state, whereas the subcritical ones are characterized by an abrupt and often localized transition leading to the coexistence of two states, one laminar and one turbulent. The results show that the transition to turbulence can occur via different regimes such as defects (amplitude) turbulence, phase turbulence or spatiotemporal intermittency, a fluctuating mixture of organized laminar domains and incoherent turbulent regions (see the article "Spatiotemporal intermittency" by H. Chaté and P. Manneville in this volume[18]). This last regime has been observed in many experiments (see below).

SOME EXPERIMENTAL CONSIDERATIONS

Whatever the exact experimental configuration, taking advantage of the quasi-1D variation of physical properties, one usually records some physical quantity $I(x,t)$. Different techniques have been developed to face this problem: moving or arrays of photodiodes[2, 3], scanning with a laser beam[1] and, more recently, systems of image processing[2, 5]. When the pattern is studied by visualization techniques, its evolution can be recorded using a video camera. The image is digitized along a line of pixels parallel to x (or along a circle in the case of an annular geometry), giving a sufficient space resolution, with the light intensity being encoded using some gray scale. The acquisition frequency has to be varied according to the intrinsic frequency f_0 of the system, and very long time series are needed to perform statistics. The spatiotemporal evolution of the system can then be reconstructed by plotting the successive lines on a space-time diagram, which is at the basis of all studies of 1D turbulence in experimental and numerical systems.

Raw data can then be Fourier transformed (along space and time) to obtain the wavenumber and the frequency of the observed dynamical regimes. More sophisticated techniques such as demodulation can be used in the case of propagating patterns to get the amplitude of the traveling waves[19]. In systems in which the transition to turbulence occurs via spatiotemporal intermittency, further reduction methods are needed to reduce the dynamics to a binary one: laminar/turbulent (see [18] and an illustration on the Rayleigh-Bénard convection experiment). A statistical analysis of the transition to turbulence can then be performed on the reduced data and different quantities such as the mean turbulent fraction, the distributions of size and lifetimes of the laminar domains and correlation lengths can be studied. Finite-size effects are present in the experiments and must not be forgotten when performing the analysis.

A MODEL SYSTEM: RAYLEIGH-BÉNARD CONVECTION

Rayleigh-Bénard convection was studied in an annular and a rectangular cell with a narrow gap in order to approach quasi-1D conditions (transverse aspect ratio $\Gamma_y \sim 0.3$), filled with silicon oil of Prandtl number 7[2]. The convective structures were visualized by shadowgraphic imaging and their spatiotemporal evolution was recorded using the technique described above. The spatial resolution was about 20 pixels per wavelength. The sampling frequency was varied between 1 image/10 sec. and 10 images/ sec. and

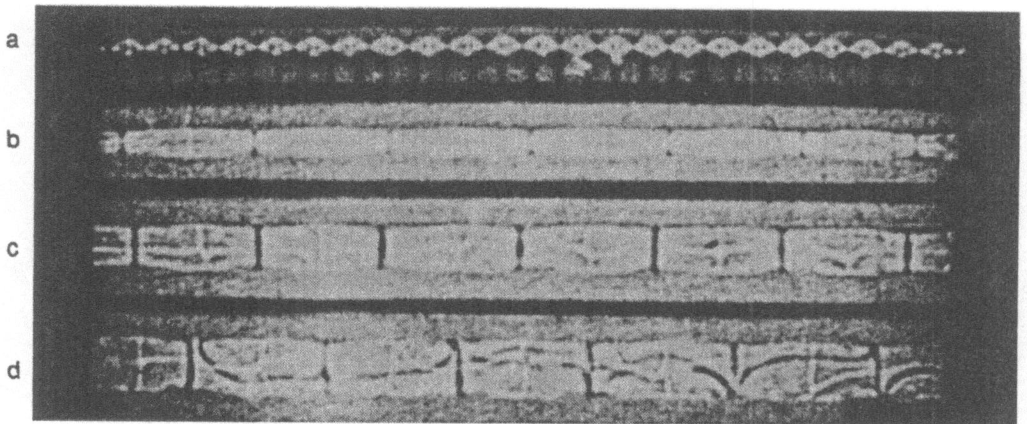

Figure 1 :   Shadowgraphic picture of Rayleigh-Bénard convection with the fourfold cell, seen from above. Bright (dark) lines correspond to cold (warm) currents. Channel (a) $\Gamma_y = 0.35$,(b) $\Gamma_y = 0.58$,(c) $\Gamma_y = 0.93$, (d) $\Gamma_y = 1.17$

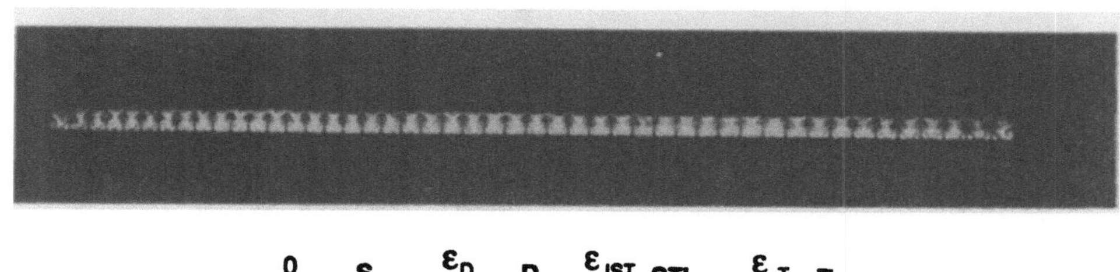

Figure 2 :   (a) Shadowgraphic image of 1D Rayleigh-Bénard convection in a rectangular container. (b) Phase diagram of the convective state as a function of the control parameter $\varepsilon = Ra/Ra_c - 1$; see text for explanations.

time series of several thousands of T_0, with $T_0 \sim 1$ sec. the basic period of oscillation, were needed to perform statistics.

The control parameter of the experiment is the Rayleigh number Ra which is proportional to the vertical temperature difference. When Ra exceeds a critical value Ra_c, a perfect stationary periodic pattern with a well defined wavelength λ_0 is observed (~ 100 rolls, cf. Fig.2a and domain labeled by S in Fig.2b). At some distance of the threshold the pattern becomes time dependent with the appearance of thermal oscillators with a period T_0 uniform in all the container (cf. Fig.3a and domain D). Further beyond threshold, turbulent patches are observed and the convective pattern enters a regime of spatiotemporal intermittency (STI domain) characterized by a mixture of organized laminar domains and incoherent turbulent zones that are fluctuating in space and time (cf. Fig.3b and c). Since turbulent patches are regions without spatial coherence and with a chaotic time evolution, different criteria can be chosen to distinguish between laminar and turbulent domains:

- a *spatial criterion:*
    if the local wavelength is such as $\lambda_0 - \Delta\lambda < \lambda < \lambda_0 + \Delta\lambda$, with λ_0 the basic wavelength of the pattern and $\Delta\lambda$ a tolerance factor, the behavior is said to be laminar, otherwise it is turbulent.

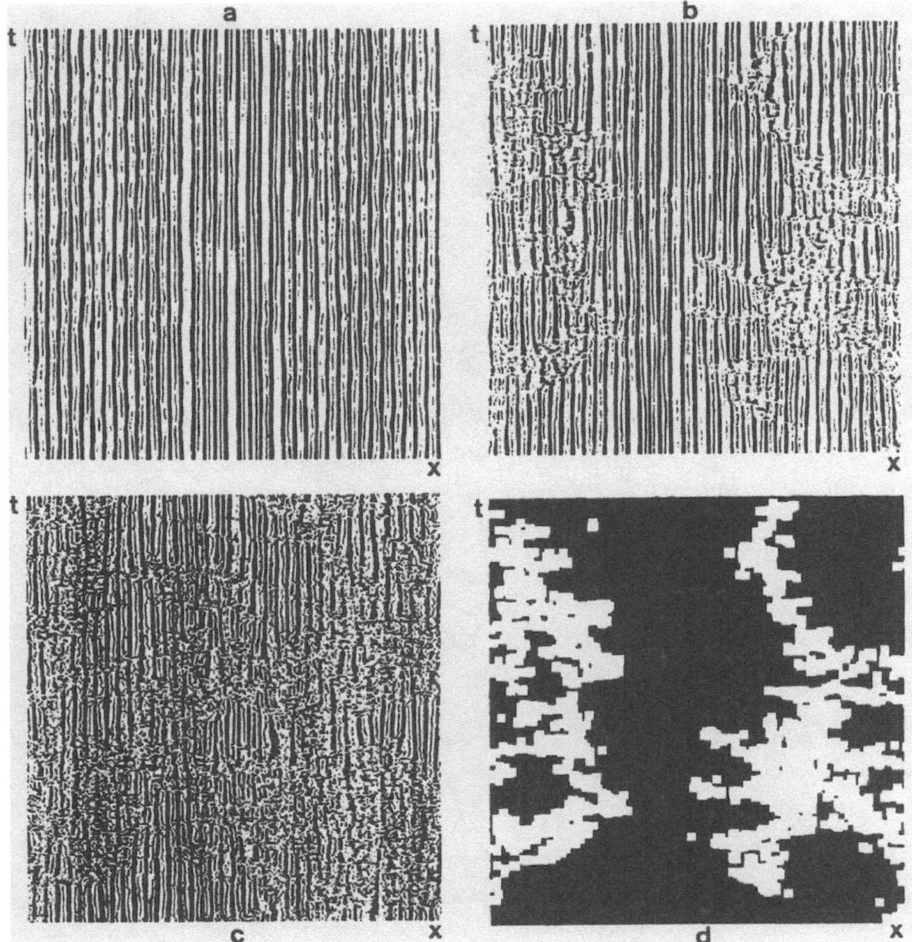

Figure 3 : Space-time evolution of the convective pattern obtained in the annular geometry for different values of the control parameter. (a) $\varepsilon = 450$; (b) $\varepsilon = 470$; (c) $\varepsilon = 500$; (d) binary reduction of (b)

- a *temporal criterion:* intensity differences between successive acquisitions can be computed and compared to a cutoff value δ. A pixel i is called laminar if: $I(i, t+1) - I(i, t) < \delta$ otherwise it is called turbulent.

One must check that such a binary reduction does not depend on the tolerance factor or the cutoff value. A more global method, based on a direct treatment of the spatiotemporal image, can also be used and give the same results (cf. Fig.3d).

At a statistical level, a global characterization of the transition is given by the variation of the turbulent fraction F_t with the control parameter: F_t is calculated as the averaged total length occupied by the turbulent cells divided by the length of the container. The variation of F_t with ε exhibits a well-defined critical behavior: $F_t \sim (\varepsilon - \varepsilon_{STI})^\beta$ with $\beta \simeq 0.3$. A power law decay is also observed at the threshold for the distribution of the number $\mathcal{N}(L)$ of laminar domains of length L and the number $\mathcal{N}(\tau)$ of laminar domains lasting τ. Above the threshold, the distributions become exponential, defining a characteristic length and time which decay as $(\varepsilon - \varepsilon_{STI})^{-0.5}$.

The experiments performed in 1D Rayleigh-Bénard convection thus reveal a route to turbulence via spatiotemporal intermittency that is analogous to a phase transition, as conjectured by Pomeau[20]. Moreover, many features are similar to these given by numerical studies, in particular those of directed percolation, even if the laminar state is only quasi-absorbing (a turbulent region may originate from a laminar one around the threshold). This experimental system thus reveals a type of STI dominated by probabilistic processes, cf. [18].

OTHER EXPERIMENTS

We now review some other quasi-1D systems and their transition to turbulence. The first two experiments reveal a transition to turbulence via spatiotemporal intermittency, as in our model case, but with some differences. The last one reveals a different kind of spatiotemporal chaos

Directional viscous fingering

Directional viscous fingering (or printer's instability) is an instability which affects the meniscus of a viscous fluid placed in the widening gap between two surfaces. The specific experimental setup considered here is composed of two off-centered cylinders which can rotate independently, one inside the other [5]. The two control parameters are the velocities V_i and V_o of the cylinders and the patterns are visualized by optical methods. When one cylinder is at rest, the unstable interface is formed by a 1D periodic pattern of about 50 steady cells (cf. Fig. 4a). When the two cylinders are rotating, the interface becomes time dependent. In the case of counter-rotation, this state is propagative, whereas, in the case of co-rotation, the system evolves to

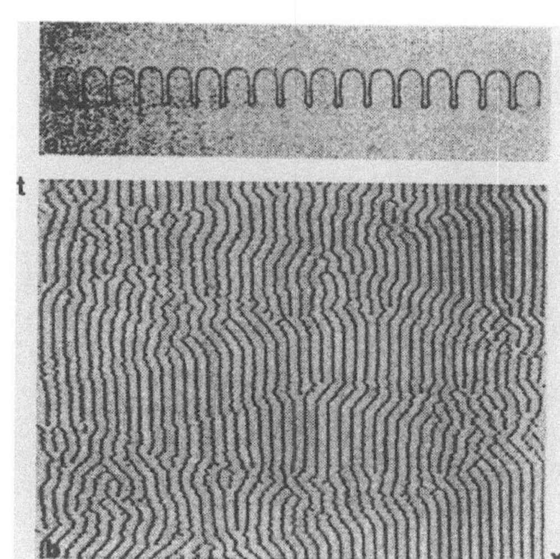

Figure 4 (a) Photograph of the meniscus above the instability threshold. (b) Space-time diagram of the spatiotemporal intermittency regime.

a chaotic dynamics characterized by a loss of spatial and temporal coherence of the cellular pattern (see Fig. 4b).

The analysis of this evolution reveals a transition via spatiotemporal intermittency, as in the convection experiments, but with some important differences both qualitative (aspects of the space-time diagrams) and quantitative (values of exponents). In this experiment, an element is a static cell and not an oscillating one. Moreover, the contamination process is dominated by the propagation and the interaction of parity-breaking waves and chaos affects the position of the cells and even their existence. In this system, the STI regime is thus dominated by deterministic, localized, propagative structures, cf. [18].

Line of vortices

In this case, the 1D pattern consists in a line of electromagnetically forced vortices obtained in a cell filled with sulfuric acid and submitted to a steady electric current which is the control parameter[12]. When the current is increased, the system undergoes a transition to a state com-

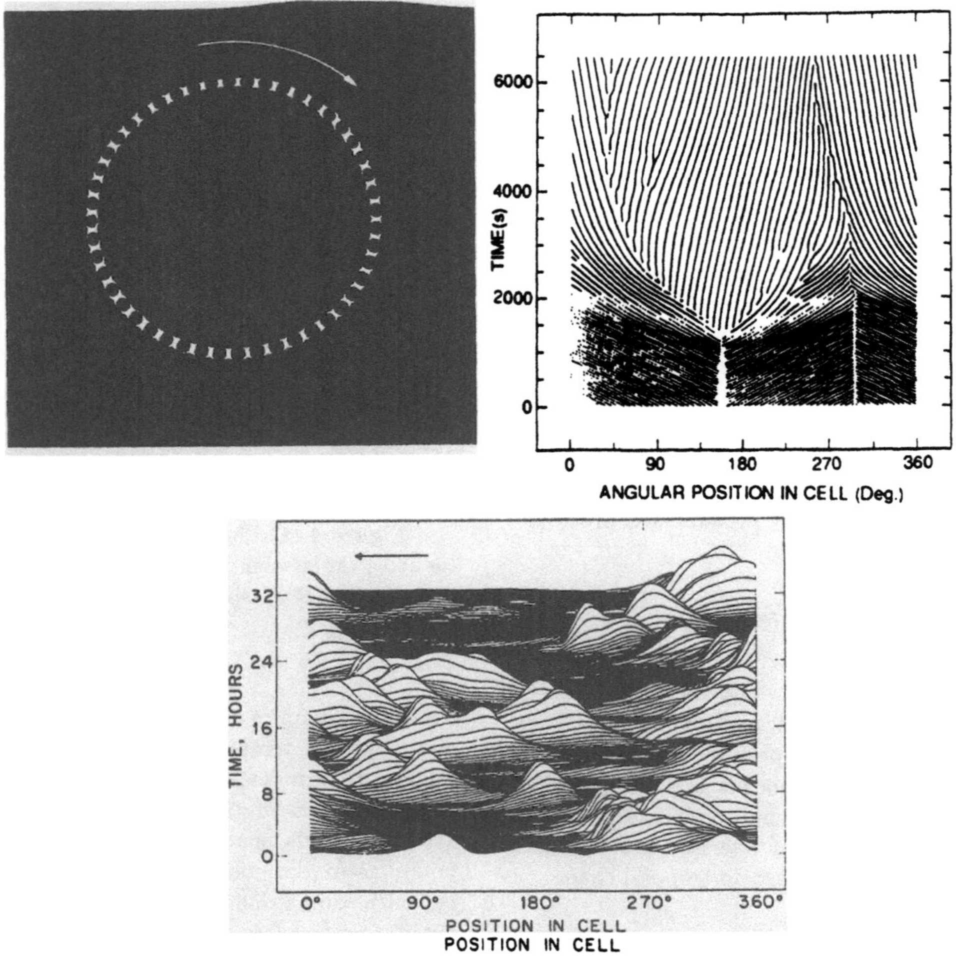

Figure 5: (a) Shadowgraph image of the traveling wave pattern in an annular container. Spatiotemporal behavior showing (b) space-time defects and (c) dispersive chaos

posed of 15 corotating vortices which are further subjected to temporal instabilities. A shadowgraphic method is used to visualize the separatrices between the vortices. For larger values of the current, the monoperiodic regime gives place to intermittent bursts, and turbulence appears via a regime of spatiotemporal intermittency. A first regime is observed, characterized by the spontaneous nucleation of turbulent domains (defined by amplitude holes and rapid phase changes). At larger values of the control parameter, the bursts connect and the system percolates. The spontaneous nucleation process of turbulent domains is dominant in this system, which thus exhibits spatiotemporal intermittency only in a loose sense, cf. [18].

Convection in binary mixtures

Rayleigh-Bénard convection in binary-fluid mixtures has emerged as a useful experimental system for exploring the spatiotemporal dynamics of nonlinear traveling waves. The 1D experimental system generally consists in a narrow, annular container filled with in an ethanol/water mixture. In this geometry, the convective rolls align radially and travel azimuthally.

The diagnostic used in these experiments is flow visualization by shadowgraphy and the control parameter is the Rayleigh number Ra. Depending on the Lewis number Le, the separation ratio ψ and on Ra, several phenomena are observed in this system. Among them, are steady and dynamical confined states, complex transients composed of spatiotemporal defects and the continuously erratic weakly nonlinear traveling wave state called dispersive chaos[3].

For large negative values of ψ and small Le, convection is triggered by the linear oscillatory instability. When Ra is increased above the threshold for the linear stability, the flow evolves, via transients, into a state of uniform overturning convection: 50 roll pairs filling the cell and propagating in a single direction (cf. Fig.5a). If Ra is then reduced below a certain threshold, a confined state of traveling rolls can be observed. The competition between these states can lead to complicated transients and spatiotemporal defects and the evolution of the flows produced can be described in terms of the stability and propagation of the fronts which separate them (cf. Fig.5b). For small negative values of ψ, nonlinear dispersion dominates the dynamics and regimes of repetitive formation and sudden collapse of spatially localized pulses are observed (cf. Fig.5c). The study of 1D convection in binary mixtures thus exhibits new routes to spatiotemporal chaos, such as amplitude and phase turbulence[21].

CONCLUSION

As can be inferred from this review, the study of 1D systems, with different physical mechanisms, reveals a very general and important route to turbulence: spatiotemporal intermittency. However, this scenario does not seem to be universal and other kinds of transitions are seen.

A transition via spatiotemporal intermittency has been observed and analyzed in 1D Rayleigh-Bénard convection, directional viscous fingering, line of vortices and seems to be present in forced convection, 1D liquid film, directional solidification and nematic liquid crystals. More theoretical work is needed to understand the differences between the two types of regimes (deterministic and probabilistic) and a new experiment showing the two behaviors would be interesting. The case of systems with traveling patterns is less clear, because they show a transition to spatiotemporal chaos by creation of defects and phase instabilities, but with no well-established scenario. This is the case of binary convection and fluid layers with a free surface locally heated. Experiments such as the printer's instability or forced Rayleigh-Bénard convection could give some insights because they show together traveling waves and spatiotemporal intermittency.

Bibliography

[1] S. Ciliberto and P. Bigazzi, Phys. Rev. Lett. **60**, 286 (1988)

[2] F. Daviaud, M. Dubois and P. Bergé, Europhys. Lett. **9**, 441 (1989); F. Daviaud, M. Bonetti and M. Dubois, Phys. Rev. A **42**, 3388 (1990); F. Daviaud, J. Lega, P. Bergé, P. Coullet and M. Dubois Physica D **55**, 287 (1992).

[3] P. Kolodner, D. Bensimon and C.M. Surko, Phys. Rev. Lett. **60**, 1723 (1988); P. Kolodner, J.A. Glazier and H. Williams, Phys. Rev. Lett. **65**, 1579 (1990); P. Kolodner, Phys. Rev. A **46**, 6431 (1992); D. Bensimon, P. Kolodner, C.M. Surko, H. Williams and V. Croquette, J. Fluid Mech. **217**, 441 (1992)

[4] S. Douady, S. Fauve and O. Thual, Europhys. Lett. **10**, 309 (1989)

[5] M. Rabaud, S. Michalland and Y. Couder, Phys. Rev. Lett. **64**, 184 (1990); S. Michalland and M. Rabaud, Physica D **61**, 197 (1992); S. Michalland, M. Rabaud and Y. Couder, Europhys. Lett. **22**, 17 (1993).

[6] L. Limat, P. Jenffer, B. Daggens, E. Touron, M. Fermigier and J.E. Wesfreid, Physica D **61**, 166 (1992)

[7] A.J. Simon, J. Bechhoeffer, A. Libchaber and P. Oswald, Phys. Rev. A **40**, 2042 (1989)

[8] G. Faivre, S. de Cheveigné, C. Guthmann and P. Kurowski, Europhys. Lett.**9**, 779 (1989); G. Faivre and J. Mergy, Phys. Rev. A **46**, 963 (1992).

[9] I. Mutabazi, J. Hegsteh, D. Andereck and J.E. Wesfreid, Phys. Rev. Lett. **64**, 1729 (1990).

[10] J.M. Vince and M. Dubois, Europhys. Lett. **20**, 505 (1992).

[11] F. Daviaud and J.M. Vince, Phys. Rev. E , (1993).

[12] H. Willaime, O. Cardoso and P. Tabeling, Phys. Rev. E **48**, 288 (1993).

[13] L. Gil, G. Balzer, P. Coullet, M. Dubois and P. Bergé, Phys. Rev. Lett. **66**, 3249 (1991).

[14] F. Daviaud, A. Burnol and O. Ronsin, Europhys. Lett. **16**, 667 (1991).

[15] P. Le Gal, C.R. Acad. Sci. Paris **313**, II, 1499 (1991).

[16] H. Chaté and P. Manneville, Phys. Rev. Lett. **58**, 112 (1987), Physica D **32**, 409 (1988), J. Stat. Phys. **56**, 357 (1988).

[17] K. Kaneko, Prog. Theor. Phys. **74**, 1033 (1985), Physica D **34**, 1 (1989).

[18] H. Chaté and P. Manneville, *Spatiotemporal intermittency*, this volume.

[19] V. Croquette and H. Williams, Physica D **37**, 300 (1989).

[20] Y.Pomeau, Physica D **23**, 3 (1986).

[21] H. Chaté and P. Manneville, *Phase Turbulence*, this volume.

[22] For a review, see e.g. M.C. Cross and P.C. Hohenberg, Rev. Mod. Phys. **65** (1993).

Experiments on 2D Turbulence
(Laboratory)

P. Tabeling
Laboratoire de Physique Statistique,
Ecole Normale Supérieure,
24, rue Lhomond, F-75231 Paris cedex 05, France

The problem of performing laboratory experiments on 2D turbulence is that we live in a three dimensional world and indeed any experiment is three-dimensional in nature. To approach a two-dimensional situation, one must inhibit the spatial degrees of freedom in a given direction and leave the others free. Since no experimental set-up achieves a complete suppression of the corresponding degrees of freedom, one should better talk of "quasi two-dimensional" situations. In all cases, some three-dimensionality enters the system, to an extent which is often delicate to determine and which is, in practice, not clearly controlled. Thus, real experiments deal with imperfect two-dimensional flows, and one of their obvious contribution is precisely to address the problem of the applicability of two-dimensional theories to real systems, including both laboratory and geophysical situations.

EXPERIMENTAL SITUATIONS

There are several methods for imposing the dynamics of a flow to be quasi two-dimensional. All of them use an extra parameter which can be a rotation, a magnetic field, a confinement, or a stratification. We shall discuss below the effect of each of them, and present examples of the corresponding experimental situations. Magnetized electron columns, whose dynamics has been shown recently to be isomorphic to that of 2D flows, are also described.

Rotation

In a rotating fluid, the parameter which controls the effect of the rotation is the Rossby number, which can be defined by:

$$R = \frac{U}{2\Omega l}$$

where U, l and Ω are typical velocity, length and rotation rate[1]. Small values of the Rossby number correspond to a large effect of the rotation. In a laboratory experiment, since U must be large enough to be in a turbulent state, the Rossby number can hardly be smaller than one but this is enough for rotation to have a significant influence on the flow. In the limit of very small Rossby numbers, the Coriolis force is dominant, and the flow organizes itself so as to balance the Coriolis force with the pressure gradient. This strong constraint is expressed by the Taylor-Proudman theorem which states that in the limit of zero Rossby number and inviscid fluids, the flow properties are invariant along the rotation axis of the system. In the remaining directions, the Coriolis forces have no dynamical effect. This is the way how rotation tends to induce two-dimensionality. The Taylor-Proudman theorem strictly applies to stationary flows. In a turbulent system, one expects that the fluctuating velocity component along the rotation axis is much weaker than the other ones, and thus that the turbulence is quasi-two-dimensional. A set-up using this mechanism consisted in a rotating container, in which the turbulence was generated, at the bottom, by an oscillating grid[2].

Close to the grid, the flow is fully three-dimensional, and above some distance, the structures become invariant along the rotation axis, revealing two-dimensionality.

Magnetic field

Another method of approaching two-dimensionality is to apply a strong magnetic field on a flow of conducting fluid. In the presence of a magnetic field, flows of conducting fluid generate electrical currents, which couple to the magnetic field to produce body forces directed against the movement[3]. This effect is controlled by the Hartmann number whose expression is:

$$M = BL\sqrt{\sigma/\mu}$$

in which B, L, σ and μ are respectively the magnetic field, a typical length, the electrical conductivity and the viscosity of the fluid. The influence of the magnetic field is large when the Hartmann number is high. In practice, it is easy to produce Hartmann numbers as large as a few hundreds in the laboratory. In the case of large Hartmann number, the flow tends to organize itself so as to cancel the induced current. This is a strong constraint, and like in the case of the rotation, one can show that, far from the boundaries, stationary flows tend to become invariant along the magnetic field as the Hartmann number tends to infinity. In the case of a turbulent flow, one expects that the magnetic field inhibits the fluctuating velocity component along the magnetic field, without affecting the other ones, so that the system also tends to be two-dimensional.

An experimental set-up for which this mechanism is relevant is composed of a container, filled with mercury, and immersed in a magnetic field, with a flow generated by electromagnetic forcing[4].

Stratification

A stable density stratification is also a method for producing quasi-two dimensional flows. The physical mechanism involved in this case is the tendency to inhibit the transfer of mass between layers of different densities, and thus to suppress a component of the velocity field. The suppression of three-dimensionality by stratification is, in some experiments[5], a spectacular phenomenon, which would certainly be interesting to investigate more deeply. In the simplest situation, after suppression of the three dimensional components, the flow develops in a thin layer, and the momentum slowly diffuses across the layer depth as time proceeds; in more complicated systems, the layer is not flat and has some dynamics (see Ref. [5]). The tendency to two-dimensionality however seems to exist even in this case. An example of such systems is a tank, linearly stratified, within which an injection of colored fluid is performed, so as to produce vortices after a short transient[5].

Confinement

A fourth method for producing quasi-two-dimensional flows consists in imposing a strong confinement in a prescribed direction so as to suppress a component of the velocity field. This is achieved in flows developing in thin layers, above an horizontal plate[6], and in soap films[7]. In thin layers, the amplitudes of the two remaining components depend on the coordinate normal to the layer; the usual assumption is to consider that the dynamics of the flow does not depend on such a coordinate, and in this sense the system is two-dimensional; to the author's knowledge, this assumption has still not been fully assessed. The situation is clearer in the case of the horizontal soap films, which has been analyzed theoretically[7]. Here both the dynamics and the structure of the flow can be considered as independent of the coordinate normal to the film, and thus the flow is incompressible and two-dimensional, provided that the velocities are smaller than the phase speed of the elastic waves. In the case where a rigid wall is present, the situation is more complex: the usual approach is to consider that the wall exerts a linear friction on the fluid, as in Hele-Shaw cells[8]. Actually, in such systems, three-dimensional recirculating flows are also present within the layer, and one may suspect that their effect does not reduce to a linear friction. To the author's knowledge, this aspect of the problem has still not been analyzed.

Magnetized electron columns

The case of magnetized electron columns is particular because it is not a two-dimensional flow of fluid particles in the usual sense, but its dynamics can be, in some circumstances, isomorphic to that of 2D flows. The correspondences are between the electron density and the vorticity, and the electrical potential and the stream function; the conservation law for the charge density is then the Euler equation of the system. This analogy can be established for a collisionless regime, over distances large compared to the gyroradius of the electron. In the experimental set-up which has been built[9], rotating columns of electrons are confined axially by an electric field, and radially by a magnetic field. As in ordinary flows, quasi-two-dimensionality results from the fact that the confinement tends to suppress the velocity fluctuations in a prescribed direction (which is parallel to the rotation axes of the columns in this case).

MEASUREMENT TECHNIQUES

The problem of the measurement is specific in quasi-two-dimensional flows. The techniques which apply for three-dimensional systems, such as hot wire anemometry, are not always transposable to quasi-two-dimensional flows: this is the case when there is no space available for the probe itself (for instance in the soap film) or, less trivially, when there is no steady mean flow (in that case, time variables cannot be converted into space variables and then no information can be obtained on the spatial structure of the flow). Another aspect of two-dimensional systems is that since they develop in a plane,

they can be fully characterized by visualization techniques; this favors the use of images to study them. These two characteristics underline the specificity of the measurement of the velocity field in quasi-two-dimensional flows. In the experiments which have been performed so far, several techniques have been used: in magnetohydrodynamics systems, the velocity has been determined locally by measuring voltage drops; in rotating systems, the velocity spectrum has been indirectly determined by considering the dispersion of a contaminant. Actually, in most of the cases, the experimentalists have designed their set-up so that a visualization of the flow on a free surface is possible. Then, by putting particles on it, or by using an interferometric method, rich information on the flow could be obtained. It has been possible, with the development of the computational facilities, to obtain quantitative information on the velocity and the vorticity fields, with a reasonable accuracy. Actually, such methods suffer from a modest spatial resolution, which explains that the study of processes like vortex filamentation or enstrophy cascade has not been performed quantitatively so far.

RESULTS

Many aspects of quasi-two-dimensional flows have been studied, such as vortex dynamics, decaying and forced two-dimensional turbulence, or equilibrium states. We shall focus here on the experimental studies of turbulent regimes.

Qualitative aspects

Remarkable aspects of freely evolving 2D turbulent flows have been observed in cylinder wakes in a soap film[10] (see Fig. 1). In such an experiment, long lived, coherent structures, involving a small number of degrees of freedom, and separated by highly disordered regions. are visible; decaying 2D turbulence thus appears as composed of islands of order embedded in a structureless sea. This picture, which

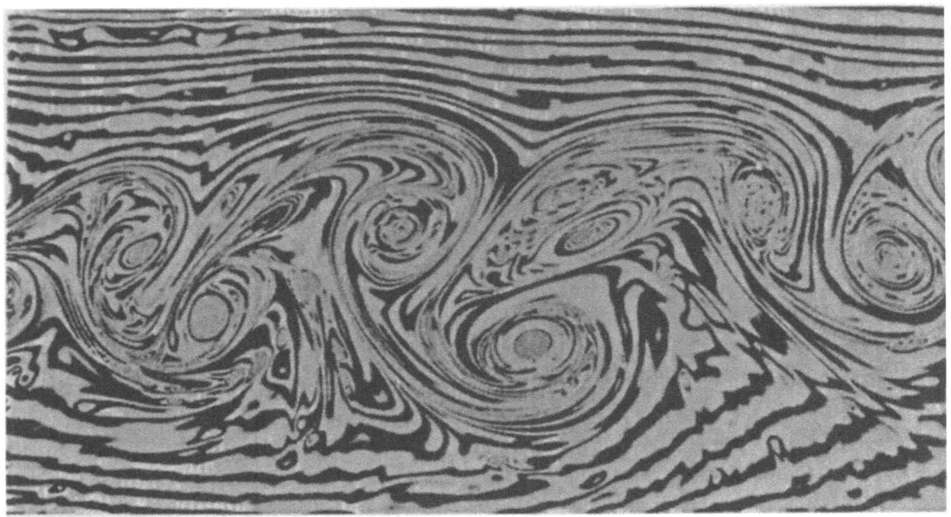

Figure 1: von Kármán street in a soap film[10]. The existence of coherent structures is obvious on this picture.

is even more evident in numerical experiments, is now commonly accepted. The situation is more confused in the case of forced turbulence, where the existence of coherent structures is not clear in laboratory experiments.

Decay of two-dimensional turbulence

The process of decay of two-dimensional turbulence has been studied in several systems, and a consistent picture of the experimental situation can be proposed. In the early experiments of Hopfinger *et al.*[2], turbulence is initially produced by an oscillating grid; then the excitation is switched off, giving rise to a decay regime. A similar procedure is used in the Cardoso *et al.*[6] experiment, with the electric current. In the case of the soap film[7], the turbulence is generated behind a comb which is driven at constant speed along the film. In all these experiments, the vortices tend to merge to form larger structures which further merge again. There is therefore strong experimental evidence for the existence of a transfer of kinetic energy towards larger scales in the decay regime. This process, which is not an energy cascade in the strict

sense since there is no constant flux of energy, roughly appears as self similar. In the experiments the transfer of energy towards large scales is limited by viscosity: as time proceeds, the mean energy of the system decreases because of viscous dissipation, so that eventually, the system ceases to be turbulent and the process stops. In practice, the largest eddy which is produced by merging can be as large as 5 times the initial ones. Measurements of the geometrical characteristics of the population of eddies during the decay phase revealed that the process is governed by power laws in time (See Fig 2). For instance, the evolution of the mean radius of the eddies $a(t)$ is governed by a power law in the form:

$$a(t) \sim t^\alpha$$

with $\alpha \simeq 0.22$ (the results in Ref. [2] have been interpreted differently, but they are consistent with this law). Other quantities such as the number of eddies, the enstrophy, etc. also decay algebraically in time and several exponents were measured. The idea of the existence of a liquid phase, where vortices play the role of atoms, was suggested to account for the high values of the vortex density and the fact that the

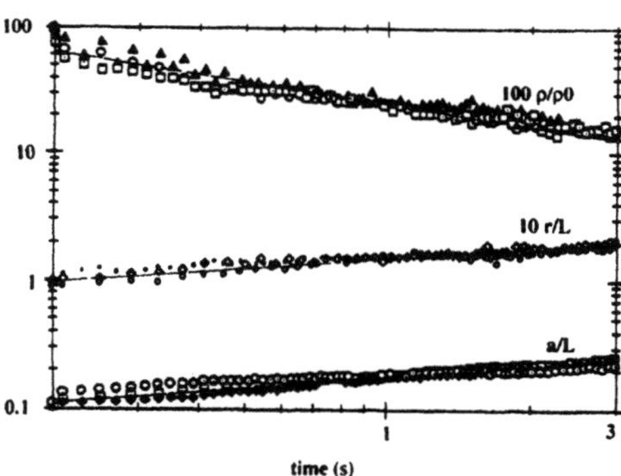

Figure 2: Temporal evolutions of the vortex densities ρ/ρ_0 (where ρ_0 is the value for $t = 0$), the mean distance between vortices $r(t)/L$, and the mean vortex sizes $a(t)/L$ (where L is the system size), for systems of 100 vortices, $b = 3mm$ and several initial values of the energy, corresponding to values of the Reynolds number about 1000. Power laws are visible, but the corresponding exponents do not agree with numerical two-dimensional estimates[12].

surface occupied by the vortices is a constant of the decay process[6].

These results are somewhat different from what we expect theoretically (see the articles by Young and Pomeau in this volume), and what has been found numerically, and up to now the origin of the discrepancy is not known. Compared with the theory, the above results agree neither with the prediction of Batchelor[11] nor with those of Carnevale *et al.*[12] in many respects. They are also different from those found in direct numerical simulations, which, incidently, are themselves controversial[13]. Thus, one can safely say that at the time where this article is written, difficulties of interpretation of the experimental results still exist.

Forced turbulence

Another topic of two-dimensional turbulence is the problem of the forced regimes. In this case, turbulence is sustained by a stationary small scale forcing, and one looks at the statistically stationary state of the flow. In the current view, the kinetic energy is not dissipated at small scales because of the absence of vortex stretching; it is transferred to larger scales, according to a cascade process. As for three-dimensional turbulence, one assumes that this cascade is characterized by a constant energy flux, which leads to an energy spectrum decreasing as the five third of the wave number. Concerning the enstrophy, it is dissipated at small scales, and the corresponding exponent of energy spectrum is expected to be comprised between 3 and 4. Several attempts have thus been made to investigate these properties in real systems; these studies have been performed by using mercury flows in the presence of a strong magnetic field[4], stably stratified fluids in a tank[5] and rotationally dominated turbulence[2]. In such experiments, the limited spatial resolution did not allow to investigate the range of scales smaller that the injection scale, so that most of the reliable results deal with the other side of the spectrum, i.e. from the injection scale up to the box size. On a qualitative level, these experiments show that the turbulent state includes a population of eddies of different sizes, extending from the injection scale up to a size comparable to a fraction of the cell. Energy spectra do reveal the existence of such a continuum of scales; actually, the spectra themselves have no remarkable feature, and it is difficult to find a region of wave numbers where a power law is clearly visible. Despite this fact, the spectra have been used, in some cases, to study the characteristics of the inverse energy cascade and provide an estimate for the two-dimensional Kolmogorov constant[4, 6]; one certainly needs more experimental work to propose a reliable value. More fundamentally, at the present time, one can say that the existence itself of an inverse energy cascade in real flows is not established.

Equilibrium states

The problem of the statistical equilibrium of two-dimensional turbulence has also been investigated experimentally. A discussion of the theoretical approach[14] is given in an article of this dictionary. One can ask if long living isolated structures, can be considered as equilibrium states of 2D turbulence. In this spirit, the internal structures of robust objects such as dipoles have been studied[5]. The general picture which emerges reveals a diversity of situations and there is no obvious way to compare these objects to isolated equilibrium structures, partly because their initial conditions and their interaction with the environment is not well defined. A better control of both the initial state and the external conditions has been achieved recently: experiments have been performed in small boxes, using electromagnetic forcing, with stratified and non stratified electrolytes; the flexibility of the system was such that several distinct initial states could be imposed. The authors showed reasonable agreement, concerning the final states, with the predictions of a maximum entropy theory in the non stratified case, whereas disagreements are observed when the working fluid is stratified[15]; this may indicate that statistical equilibrium is not reached in two-dimensional systems, whereas it can be achieved when some three-dimensional effects are at work. Actually, in another recent study, performed in a magnetized electron column[9], the best agreement is found with another theory, based on the existence of a minimum enstrophy principle[16]. One must also mention an experiment[17], performed in mercury, which tends to support the statistical analysis of Ref. [14]. The situation is therefore somewhat puzzling and certainly needs further clarification. One difficulty for carrying out this task is to find a physical situation where the predictions of the various statistical approaches strongly differ from each other. In the case of the plasma experiment[9], the differences between them did not exceed 20%. We may need, in this domain, study directly the statistical fluctuations to draw out accurate conclusions, because it is not clear, after all, that in two-dimensional system, there exist fluctuations which play a role

analogous, in any sense, to thermal noise. We refer the reader to the article of Y. Pomeau, in this volume, for a discussion of this problem.

CONCLUSION

The main conclusion is that, concerning the laboratory experiments on two-dimensional turbulence, the situation is rich, complex, and somewhat puzzling. Some ingredients of two-dimensional turbulence are obviously present in the laboratory experiments: this is the case of the coherent structures and the process of vortex merging. However, several important aspects of these systems are found different from what we expect theoretically, or from numerical simulations: this is the case for the laws of the decay regime, the existence of the inverse cascade, and the structure of the equilibrium states. All this probably merits further investigation. One could safely say that in the domain of two-dimensional turbulence, the role of laboratory experiments is not restricted to checking theoretical ideas, but also to raising new questions.

Bibliography

[1] Batchelor, G.K., An Introduction to Fluid Mechanics (Cambridge University Press) ; see also Hopfinger, E., Van Heijst, G., Ann. Rev. Fluid. Mech. **25**, 241 (1993).

[2] Hopfinger, E., Griffiths, R.W., Mory, M., Jour. Mec. Theor. Appl., n Special **21** (1983).

[3] Shercliff, A Textbook of Magnetohydrodynamics (Cambridge University Press)

[4] Sommeria, J., J. Fluid Mech., **170**, 139 (1986).

[5] Van Heijst, G. and Flor, J., Letters to Nature **340**, 212 (1989), Narimousa, S. Maxworthy, T., Spedding, G., J. Fluid. Mech. **223**, 113 (1991).

[6] Cardoso, O., Marteau D., Tabeling P., Phys. Rev. E **49**, 1 (1994).

[7] Chomaz, J.M., Cathalau, B., Phys. Rev. A **41**, 2243 (1990).

[8] Dolzhanskii F., Krymov, V., Manin, D., J. Fluid Mech **241**, 705 (1992).

[9] Fine, K., Driscoll, C., Malmberg, J., Mitchell, T., Phys. Rev. Lett. **67**, 588 (1991) ; Huang, X., Ph.D. U. San Diego (1993).

[10] Couder, Y., C.R.A.S., **297**, 641 (1983), J. Phys. Lett. 45, 353 (1984).

[11] Batchelor, G.K., Phys. Fluids Suppl. II **12**, 233 (1969).

[12] Carnevale, G.F., McWilliams, J.C., Pomeau, Y., Weiss, J.B., Young, W.R., Phys. Rev. Lett. **66**, 2735 (1991).

[13] Dritschel D., Phys Fluids A March (1993).

[14] Onsager, L., Nuevo Cim., II, **6**, 279 (1949), Miller, J., Phys. Rev. Lett., **65**, 2137 (1990), Robert, R., Sommeria J., J. Fluid. Mech. **229**, 291 (1991).

[15] Marteau, D., Cardoso, O., Tabeling, P., Preprint LPS, submitted to Phys. Fluids (1994).

[16] Leith, C., Phys. Fluids **27**, 1388 (1984).

[17] Denoix, M., Thèse INPG Grenoble (1992).

Experiments on Spatiotemporal Chaos
(in Two Dimensions)

J.P. Gollub
Haverford College,
Haverford, PA 19041, USA
and University of Pennsylvania,
Philadelphia, PA 19104, USA

A variety of experiments on spatiotemporal chaos in two dimensions are summarized. Despite a decade of work, an adequate conceptual framework and set of methods for characterizing the phenomena have not yet emerged.

INTRODUCTION

After the extensive development of the field of nonlinear dynamics in the 1980's, it became clear that the methods of low-dimensional dynamical systems could not be easily generalized to systems with many degrees of freedom. The problem of turbulence in fluids has especially been difficult to attack because of its wide range of scales.

Many investigators have turned their attention to the problem of *spatiotemporal chaos*, which is thought to be inherently simpler. As for turbulence, our concern here is with fluctuating systems in which the range ξ of spatial correlations is small compared to the size of the system. However, in *contrast* to the case of turbulence, systems exhibiting spatiotemporal chaos do not manifest a wide range of excited scales. Rather, they consist of fluctuating regions of comparable size that interact with each other on a relatively slow time scale. Situations of this type have been found to arise especially where one spatial dimension is strongly constrained geometrically, e.g., in thin layers.

Indeed, a major body of experimental literature has appeared on a great variety of systems manifesting spatiotemporal chaos in two dimensions, including parametrically excited surface waves[1-4], Rayleigh-Bénard convection[5-8], rotating convection[9, 10], binary convecting mixtures[11, 12], electro-convection in nematic liquid crystals[13-15], and film flows[16, 17]. These references are intended to be only a sampling of the extensive work that has been reported.

In this brief review I briefly summarize the observations that have been made on spatiotemporal chaos in some of these systems. The discussion is limited to two-dimensional systems, basically thin layers that are homogeneous with respect to spatial translations. Spatiotemporal dynamics that occurs predominantly in *one* spatial dimension, which has been studied for example in the printer's instability and in annular thermal convection, is treated in another contribution to this volume.

PARAMETRICALLY EXCITED SURFACE WAVES

Parametrically excited surface waves are produced via the Faraday instability when a container is subjected to a sufficiently large vertical vibration. Several different ordered patterns can be obtained above the threshold driving amplitude A_c, including squares, lines, hexagons and triangles, depending on parameters and the nature of the driving (single or multi frequency). The transition to spatiotemporal chaos has been studied in detail for the square pattern, which occurs at low viscosity and consists approximately of a superposition of standing waves in two orthogonal directions. The boundaries of the container, if significant, can orient the pattern and induce defects.

As the driving amplitude is increased, a well-defined transition to disordered waves is found for a moderately large square container, as reported by Tufillaro, Ramshankar, and Gollub[1, 2]. The threshold excitation amplitude A_t, expressed as $\epsilon_t = (A_t - A_c)/A_c$, depends on the fluid, and is much less than 1 for several cases (n-butanol and water) that have been studied in detail. The transition is mediated by a long wavelength secondary instability that has been elucidated experimentally by Ezersky *et al.*[4] and theoretically by Milner[18]. At the transition the translational correlation length falls precipitously, from a value comparable to the dimension of the system, to a much smaller value comparable to the wavelength. Orientational correlations also decline sharply at the transition, but significant correlations persist to much higher ϵ. The time scale of fluctuations, measured locally, appears to diverge as the transition is approached from above. The light intensity along an arbitrary direction in the image plane was also studied by time-resolved spatial Fourier analysis. The probability distribution of individual fluctuating spectral components was found to be roughly Gaussian in the chaotic regime, as might be expected for uncorrelated motion.

The transition process is known to be dependent on the fluid viscosity. At low viscosity, the band of unstable wavenumbers is very small, as pointed out by Edwards and Fauve[19]; as a consequence, the patterns are somewhat sensitive to container size and shape even when the size is large compared to the wavelength. There is some hope that the opposite case of large viscosity will be independent of boundaries. Though some observations have been made, this case has not yet been studied in detail.

Other transitional scenarios can also occur[4, 20]. For example, Bosch and van de Water[3] studied spatiotemporal chaos in silicon oil. In this case, the square pattern becomes time-dependent through nucleation of defects (rather than a secondary instability), and gives way to a time-dependent triangular one before the fully chaotic regime is reached. It is not known whether the modestly higher viscosity is primarily responsible for the somewhat different transitional behavior than that observed by Tufillaro, Ramshankar, and Gollub. Bosch and van de Water measured the mutual information function as a way of detecting spatial correlations. They noted that correlations fall off algebraically with distance, and in that sense may be said to have a long range character even in the chaotic regime. They also found that small patches of a definite symmetry but random orientation appear in the fluctuations; these fluctuating patches apparently cause the local probability distribution of the measured light intensity to deviate significantly from a Gaussian function in the tails. However, caution should be used in interpreting measurements of probability distributions here and in Ref.[2] because the measured light intensity is a nonlinear function of the surface wave deformation.

Disordered Faraday waves also provide a fascinating context in which to study transport and mixing phenomena. It is well known that the trajectories of fluid elements can be much more complicated than the Eulerian velocity field. Particles placed on the surface of a disordered capillary wave field undergo a random walk that is well described as a normal Brownian process if the waves are sufficiently disordered. If the surface is marked with a spot of dye, the isoconcentration contours evolve into fractal curves (prior to the eventual homogenization)[21, 22]. This phenomenon has given rise to quite a few efforts at theoretical modeling.

Gluckman *et al.*[23] have studied the time average properties of fluctuating Faraday waves. They discovered that chaotically evolving patterns can have highly ordered time averages in containers that are considerably larger than the correlation length. The instantaneous patterns are quite disordered, fluctuate rapidly, and have a correlation length (for fluctuations relative to the average pattern) that is considerably shorter than the cell dimension. On the other hand, the statistical average over a sufficiently long time is strikingly ordered.

Gluckman *et al.* found that the geometry of the averaged patterns is related to the spatial symmetry of the boundaries. In the circular case the instantaneous patterns have a tendency toward locally perpendicular alignment with respect to the boundaries, but the time averaged pattern has rings parallel to the boundary. The mean wavenumber in both geometries shows a statistical quantization effect, with jumps as a function of the driving frequency. The structured averages imply a remarkable phase rigidity of the chaotic state.

The amplitude of the ordered average is dependent on the lateral boundary conditions, being stronger when the boundary conditions "pin" the phase of the wave pattern. The structured average disappears at sufficiently high excitation amplitude, and for sufficiently large aspect ratio.

The time-dependent fluctuations relative to the average pattern reveal interesting collective effects. For example, Gluckman *et al.*[23] noted short lived patches that are regular and spatially in phase with the average pattern. This phase rigidity is responsible for the structured average patterns. The fluctuating patches noted in this experiment are presumably related to those described by Bosch and van de Water[3]. This latter group has confirmed the existence of structured average wave patterns, and has in addition shown that the wave maxima move erratically around their average positions in much the same way as do the atoms in a thermally excited two-dimensional lattice[24].

RAYLEIGH-BÉNARD CONVECTION: PURE FLUIDS

Experiments aimed at understanding the onset of spatiotemporal disorder have focused more extensively on Rayleigh-Bénard convection than on any other system. There are several striking recent examples. Morris *et al.*[8] have succeeded in obtaining a uniform convecting layer of much larger aspect ratio than those studied previously. They find that a chaotic pattern occurs consisting of rotating spirals and other defects. The time scale for fluctuations of the pattern decreases as a function of the control parameter ϵ defined as the normalized distance above the convective onset. For small ϵ, the patterns are time-dependent, but the curvature is rather small, and the patterns are boundary dominated. For larger ϵ, the rotating spirals are clearly independent of the boundaries.

The patterns have been characterized by their correlation length ξ, which is found to have a power law decay as a function of ϵ. Furthermore, the wavenumber band specified by ξ^{-1} is well within the range of wavenumbers for which linear stability theory predicts stable rolls. This very significant disagreement with theory remains to be resolved. The authors speculate that the stable roll state may exist, but requires special initial and boundary conditions to be realized at large aspect ratios.

Experiments on rotating Rayleigh-Bénard convection also show clear examples of spatiotemporal chaos[9, 10]. Rolls are unstable in the presence of rotation as a result of the Küppers-Lortz instability, which causes roll patches to become reoriented frequently. The resulting patterns contain a large concentration of defects (mainly dislocations and disclinations). Despite the pronounced spatiotemporal chaos, the patterns in moderate aspect ratio cells were found to have structured time averages by Ning, Hu and Ecke[25], much as occurs in the Faraday experiments.

CONVECTING BINARY MIXTURES

Convecting binary mixtures show patterns that are inherently time-dependent near onset with respect to patterns composed of traveling waves. Spatiotemporal chaos is included among the many phenomena that can arise, as described by Steinberg *et al.*[12]. In the early experiments it appeared that the chaotic state depended on having a cell that was spatially extended in two directions; the resulting dynamics is fairly complex, and a large aspect ratio study has not yet been performed in two dimensions. More recently, Kolodner, Glazier, and Williams[11] have discovered and carefully characterized a one-dimensional form of chaos involving dramatic fluctuations in the amplitudes of the traveling waves that form the pattern.

ELECTROCONVECTING NEMATICS

A thin layer of nematic liquid crystal may be driven by an AC electric field into a convecting cellular state consisting of rolls that are much like those in Rayleigh-Bénard convection. Under certain conditions the rolls can be traveling rather than stationary. As the electric field is increased, a transition to spatiotemporal chaos occurs by nucleation of pairs of topological defects in the roll pattern[13, 14]. The defects appear to arise from a phase instability of the underlying pattern. Once created, the defects propagate through the pattern, interact, and annihilate. A statistical equilibrium is reached in which a certain mean density of defect pairs is present at a given time. A rough parallel has been drawn between this behavior and that exhibited numerically in simulations of the complex Ginzburg-Landau equation[26]. However, the actual mechanism of the transition appears to depend on parameters. A somewhat different process for creating spatiotemporal chaos in electroconvecting nematics has been discussed by Nasuno and Sawada[15]. In this process phase waves (traveling disturbances in the phase of the pattern) are emitted from localized but stationary sources within the pattern.

FILM FLOWS

In all of the above examples, the primary pattern-producing instability is of the absolute type. There is also a substantial body of work on open flow systems having so-called "convective" instabilities (no relation to Rayleigh-Bénard convection), in which the disturbances grow only in a moving frame of reference, and are sustained by noise in the laboratory frame. A good example that has been extensively studied recently arises in the flow of a thin film down an incline[16, 17]. The basic flow is unstable with respect to the formation of interfacial waves. However, even if these waves are periodically forced to ensure an initial periodicity, they are in turn unstable with respect to a variety of two- and three-dimensional instabilities. The two-dimensional instabilities include sideband and subharmonic instabilities; the three-dimensional ones lead to transverse modulations of the traveling waves. At a sufficient distance downstream, a disordered state is reached that is dominated by solitary wave bursts at irregular intervals. Statistical studies of this type of spatiotemporal chaos are under way.

DISCUSSION AND CONCLUSION

As the reader can see from this limited survey, two-dimensional spatiotemporal chaos is widespread; many examples have been discovered and studied experimentally. In parallel with the experimental work, there has also been a considerable numerical and analytical effort, much of it directed at model equations such as the complex Ginzburg-Landau equation, which is believed to provide a plausible approximation to the physics for some systems in which spatiotemporal chaos occurs close to the onset of the basic pattern. The Kuramoto-Sivashinsky equation has provided another approach to modeling, as have coupled map lattices and other discrete models. The theoretical work is now so extensive and diverse that it is impossible to reference adequately; it is thoroughly discussed in a lengthy review by Cross and Hohenberg[27].

A number of fairly well defined phenomena have been elucidated in the theoretical literature, including the following: *spatiotemporal intermittency*, in which "laminar" (or ordered) and "chaotic" (or disordered) domains coexist; *phase turbulence*, in which the pattern amplitude is relatively constant and its spatial phase fluctuates; and *defect mediated "turbulence"* in which the pattern irregularities are concentrated in localized defects, which are nucleated and annihilated at irregular intervals and locations. The initial bifurcations leading to these states have also been investigated in some model systems.

However, many of the experimental phenomena do not fit very well into these categories. Despite the large amount of effort devoted to the problem of spatiotemporal chaos, we still do not have an adequate phenomenology that would allow us to understand the similarities and differences observed in the various cases.

Although the term "chaos" is frequently utilized in connection with deterministic nonlinear phenomena in extended systems, the standard methods of low-dimensional nonlinear dynamics are very difficult to apply experimentally here. These include attempts to reconstruct attractors or to measure the Lyapunov exponents describing the linearized dynamics. Proposed generalizations to extended systems have been reviewed by Abarbanel *et al.*[27].

Moreover, there is no agreement on methods for characterizing and distinguishing the different forms of spatiotemporal chaos. Though statistical methods are obviously needed, some of the more obvious candidates are inadequate. For example, it is clear that correlation functions do not discriminate adequately among the different forms of disorder and do not reveal the coherent structures that are visible to the eye. Clearly, this subject is still as murky as the broader problem of turbulence.

ACKNOWLEDGMENTS

The financial support of the National Science Foundation under Grants CTS-9115005 and DMR-9319973 is gratefully acknowledged.

Bibliography

[1] J.P. Gollub and R. Ramshankar, "Spatiotemporal chaos in interfacial waves," in New Perspectives in Turbulence, L. Sirovich, Eds. (Springer-Verlag, New York, 1991), p. 165.

[2] N.B. Tufillaro, R. Ramshankar and J.P. Gollub, "Order-disorder transition in capillary ripples," Phys. Rev. Lett. **62**, 422 (1989).

[3] E. Bosch and W. van de Water, "Spatiotemporal intermittency in the Faraday experiment," Phys. Rev. Lett. **70**, 3420 (1993).

[4] A.B. Ezerskii, M.I. Rabinovich, V.P. Reutov and I.M. Starobinets, "Spatiotemporal chaos in the parametric excitation of a capillary ripple," Sov. Phys. JETP **64**, 1228 (1986).

[5] G. Ahlers and R.P. Behringer, "The Rayleigh-Bénard instability and the evolution of turbulence," Prog. Theor. Phys. Supp. **64**, 186 (1978).

[6] M.S. Heutmaker and J.P. Gollub, "Wavenumber distributions and time dependence in Rayleigh-Bénard convection" Physica D **23**, 230 (1986).

[7] V. Croquette, "Convective pattern dynamics at low Prandtl number," Contemp. Phys. **30**, 113 (1989).

[8] S.W. Morris, E. Bodenschatz, D.S. Cannell and G. Ahlers, "Spiral defect chaos in large aspect ratio Rayleigh-Bénard convection," Phys. Rev. Lett. **13**, 2026 (1993).

[9] L. Ning and R.E. Ecke, "Küppers-Lortz transition at high dimensionless rotation rates in rotating Rayleigh-Bénard convection," Phys. Rev. E **47**, 2991 (1993).

[10] F. Zhong, R. Ecke and V. Steinberg, "Rotating Rayleigh-Bénard convection: The Küppers Lortz transition," Physica D **51**, 596 (1991).

[11] P.Kolodner, J.A. Glazier and H. Williams, "Dispersive chaos in one-dimensional traveling-wave convection" Phys. Rev. Lett. **65**, 1579 (1990).

[12] V. Steinberg, J. Fineberg, E. Moses and I. Rehberg, "Pattern selection and transition to turbulence in propagating waves," Physica D **37**, 359 (1989).

[13] I. Rehberg, S. Rasenat and V. Steinberg, "Traveling waves and defect-initiated turbulence in electroconvecting nematics," Phys. Rev. Lett. **62**, 756 (1989).

[14] E. Braun, S. Rasenat and V. Steinberg, "The mechanism of transition to weak turbulence in extended anisotropic systems," Europhys. Lett. **15**, 597 (1991).

[15] S. Nasuno and Y. Sawada, "A new scheme of spatio-temporal chaos created by the interaction between phase waves and topological defects," Progr. Theor. Phys. Suppl. **99**, 450 (1989).

[16] J. Liu, J.D. Paul and J.P. Gollub, "Measurements of the primary instabilities of film flows," J. Fluid Mech. **250**, 69 (1993).

[17] J. Liu and J.P. Gollub, "Onset of spatially chaotic waves on flowing films," Phys. Rev. Lett. **70**, 2289 (1993).

[18] S.T. Milner, "Square patterns and secondary instabilities in driven capillary waves," J. Fluid Mech. **225**, 81 (1991).

[19] W.S. Edwards and S. Fauve, "Patterns and quasipatterns in the Faraday experiment" J. Fluid. Mech. (in press).

[20] M.I. Rabinovich and M.M. Sushchik, "The regular and chaotic dynamics of structures in fluid flows," Sov. Phys. Usp. **33**, 1 (1990).

[21] R. Ramshankar and J.P. Gollub, "Transport by Capillary Waves. Part I: Particle Trajectories," Phys. Fluids A **2**, 1955 (1990).

[22] R. Ramshankar and J.P. Gollub, "Transport by Capillary Waves. Part II: Scalar Dispersion and Structure of the Concentration Field," Phys. Fluids A **3**, 1344 (1991).

[23] B.J. Gluckman, P. Marcq, J. Bridger and J.P. Gollub, "Time-averaging of chaotic spatiotemporal wave patterns," Phys. Rev. Lett. **71**, 2034 (1993).

[24] E. Bosch, H. Lambermont and W. van de Water, "Average patterns in the Faraday effect" Phys. Rev. E **49**, R3580 (1994).

[25] L. Ning, Y. Hu and R.E. Ecke, "Spatial and temporal averages in chaotic patterns," Phys. Rev. Lett. **71**, 2216 (1993).

[26] P. Coullet, L. Gil and J. Lega, "A form of turbulence associated with defects," Physica D **37**, 91 (1991).

[27] M.C. Cross and P.C. Hohenberg, "Pattern formation outside of equilibrium," Rev. Mod. Phys. **65**, 851 (1993).

[28] H.D.I. Abarbanel, R. Brown, J.J. Sidorowich and L.S. Tsimring, "The analysis of observed chaotic data in physical systems," Rev. Mod. Phys. **65**, 1331 (1993).

EXTENDED SELF SIMILARITY

S. CILIBERTO

Laboratoire de Physique, CNRS URA 1325
Ecole Normale Supérieure,
46 Allée d'Italie, F-69364 Lyon, France

INTRODUCTION

In fully developed three dimensional turbulence the name extended self similarity (E.S.S.) is used to indicate a special property of the n-order structure functions $< \Delta V(r)^n >$, where $\Delta V(r) = V(x+r) - V(x)$ and $V(x)$ is the velocity component at the position x parallel to the relative displacement r and $< .. >$ means space average. Specifically it has been found [1, 2] that:

$$< \Delta V(r)^n >= C_n U^n [\frac{r}{L} f(r/\eta)]^{\zeta(n)} \quad (1)$$

where U is a characteristic velocity, η is the Kolmogorov scale and L the integral scale. The function $f(r/\eta)$ is an universal function that can be computed[1] using the Kolmogorov equation[3]:

$$< \Delta V(r)^3 > = -\frac{4}{5}\epsilon\, r + 6\nu\frac{\partial\, <\Delta V(r)^2>}{\partial r} \quad (2)$$

where ϵ is the energy dissipation and ν the kinematic viscosity. For r in the inertial range ($\eta << r << L$, that is when the second term in eq.(2) may be neglected) eq.(1) reduces to the standard scaling form of structure functions[3]:

$$< \Delta V(r)^n > \approx r^{\zeta(n)} \quad (3)$$

From eq.(2) one immediately sees that $\zeta(3) = 1$. For the other values of n the theoretical expectations[3] based on the Kolmogorov 1941 theory is that $\zeta(n) = \frac{n}{3}$. On the other hand experiments[4] and recent numerical simulations indicates that $\zeta(n)$ is a non linear function of n. However we

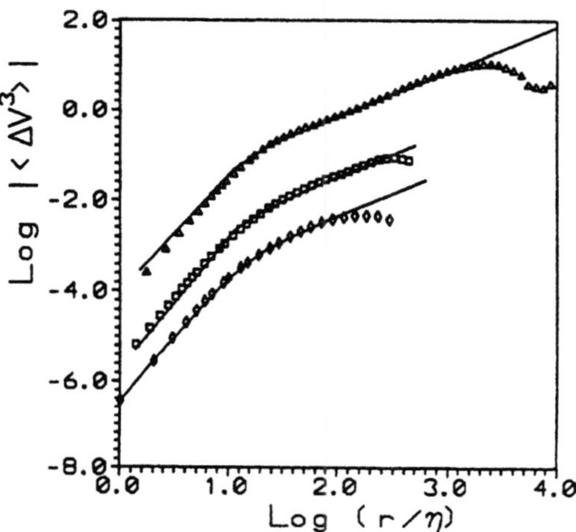

Figure 1: Third order structure function for $Re = 3 \cdot 10^5 \triangle$, $Re = 47000 \,\square$, $Re = 6000 \,\diamond$

remind that eq.(3) holds only at very large Reynolds number(Re), where clear scaling laws are observed. In contrast the scaling of eq.(1), that is E.S.S., holds with the same $\zeta(n)$ at high as low Re and extends almost to the Kolmogorov scale η. These properties of eq.(1) allow to achieve a much better estimate of $\zeta(n)$, because eq.(1) is a consequence of the self scaling properties of the velocity structure functions[1, 2]:

$$< \Delta V(r)^n > = A_n | < \Delta V(r)^3 > |^{\zeta(n)}$$
$$= B_n < |\Delta V(r)^3| >^{\zeta(n)} \quad (4)$$

which has been successfully used to estimate $\zeta(n)$ at very low Re[1, 2]. A_n and B_n are two different sets of constant and the relation

$$| < \Delta V(r)^3 > | = B_3/A_3 < |\Delta V(r)^3| > \quad (5)$$

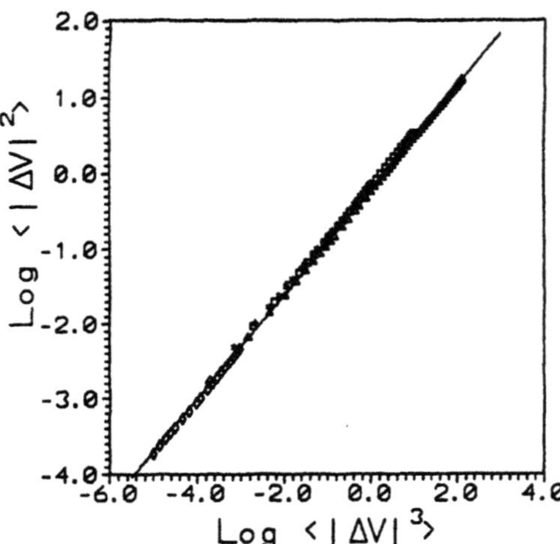

$$\text{Log} \langle |\Delta V|^2 \rangle \text{ vs } \text{Log} \langle |\Delta V|^3 \rangle$$

Figure 2: Second order structure function as a function of the third one for $4000 < Re < 3 \cdot 10^5$

which cannot be deduced by Navier-Stokes is verified experimentally[2]. The reason for the use of $T(r) = \langle |\Delta V(r)^3| \rangle$ is that is statistically more stable than

$$S(r) = | \langle \Delta V(r)^3 \rangle |$$

At the moment there is no theoretical justification for eq.(1) and (4) but very strong experimental[1, 2] and numerical[5] evidence, in the case of isotropic turbulence (far from boundaries), for $10^3 < Re < 10^6$. Thus we will summarize here the experimental results which lead to write eq.(1). It is important to stress that an equation similar to eq.(1) hold also for the structure functions of passive scalars. Furthermore experimental and numerical data show that E.S.S. can be successfully applied to turbulent thermal convection[6, 7].

EXPERIMENTAL EVIDENCE OF E.S.S.

Here we will briefly summarize only the case of the velocity field in homogeneous turbulence using experimental data obtained by hot wire measurements in a wind tunnel covering the range $10^3 < Re < 10^6$. Turbulence has been produced using many

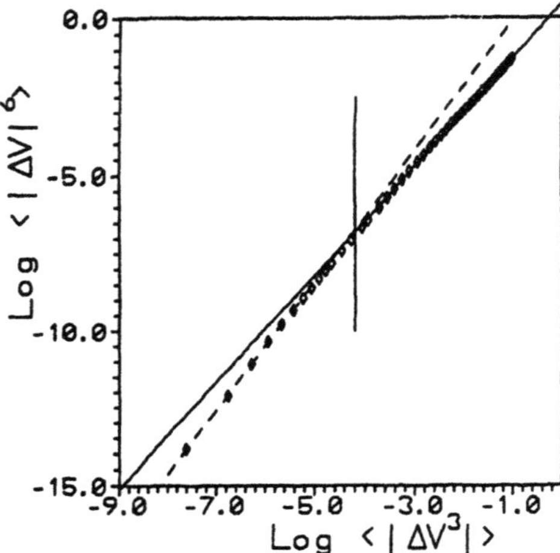

Figure 3: Six order structure function as a function of the third one for $Re = 6000$. The vertical line corresponds to $T(5\eta)$. The dashed line is a best fit with slope 2 for $T(r) < T(\eta)$. The continuous straight line has a slope $\zeta(2) = 1.75$

different flows (grid turbulence, wakes behind a cylinder and jet) and the description of the experimental apparatus can be found in ref. [8].

In figure 1 we show the third order structure function against $\frac{r}{\eta}$. Only for the two measurements at highest Re a clear inertial range ($S(r) \propto r$) is detectable. Whereas at $Re = 6000$ the inertial range is just a few points. In figure 2) we show the self scaling of $\langle \Delta V(r)^2 \rangle$ versus $T(r)$ for many different Re. We see that all data, even those at $Re = 4000$, are on a straight line of slope $\zeta(2) = 0.700$, consistent with previous estimated reported in literature. This result shows that E.S.S., in the form of the self scaling (4), allows a very accurate evaluation of $\zeta(n)$ even at very small Re.

To discuss the range of validity of E.S.S. we show, in fig.3, $\Delta V(r)^6$ versus $T(r)$ for the case at $Re = 6000$ shown in fig.1. At variance with figure 1 the self scaling (4) holds for a much wider range of scales almost down to the Kolmogorov scale. Indeed the vertical continuous line of fig.2b indicates the value $T(5\eta)$. The best fit represented by the continuous line in figure 3) gives a scaling exponent $\zeta(6) = 1.75$, i.e. exactly the same value of the scaling exponent observed at high Re, in our experiment and to the value reported in literature estimated with the standard scaling of eq.(3)[4]. At small scale, i.e. for r smaller than η, the scaling is changed and for r small enough we find (dashed straight line) a local exponent very close to 2, i.e. very close to the one obtained by the fact that the flow is regular for $r < \eta$[3]. We can show now that E.S.S. is consistent with eq.(2). Because the self scaling (4) is valid till $r \simeq \eta$, one can use eq.(4) in eq.(2) to obtain

$$-S(r) = -4/5 \ \epsilon r + 6\nu A_2 \ \frac{d}{dr} S(r)^{\zeta(2)} \quad (6)$$

We integrated eq.(6) with the parameter corresponding to the data reported in fig.1 and 2. We estimated A_2 and $\epsilon = (5/4) \cdot [S(r)/r]$, for r in the inertial range. We use $\zeta(2) = 0.7$ which is equal for all Re. In fig.1a we compare the numerical solution of eq.(6)(continuous line) with the experimental results. The agreement is excellent taking into account that no adjustable parameter has been used.

Finally we show in fig.4 the form of the function $f(r/\eta)$ which has been computed in the following way. Once $\zeta(n)$ is estimated using eq.(4) one obtains

$$f(r/\eta) = \frac{< \Delta V(r)^n >^{1/\zeta(n)} \ L}{r D_n}$$

where L/D_n is evaluated by imposing $f(r/\eta) = 1$ for $r \simeq L$. We see that doing in this way the $f(r/\eta)$, evaluated from structure functions of different order n and obtained from measurements at various Re, collapse on the same curve showing the exactitude of eq.(1). The curves begin to deviate at $r \simeq \eta$.

CONCLUSIONS

To summarize the results shown so far we can say the following: the self scaling behavior of the structure functions holds

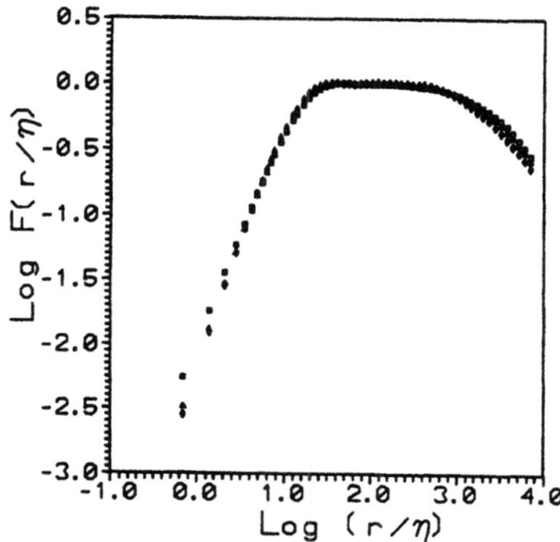

Figure 4: Universal functions $f(r/\eta)$ for different n and Re.

on a much wider range of scales with respect to the inertial range and for much lower Reynolds number with respect to the common belief. We want to remark a possible source of errors in applying self scaling (4). Scaling law (4) should be observed for fully developed turbulence far from boundaries, i.e. where the velocity field is locally isotropic. Velocity measurements taken too close to boundary layers might destroy the scaling (4). This problem is under current investigation.

Bibliography

[1] R. Benzi, S. Ciliberto, R. Tripiccione, C. Baudet, S. Succi, Phys. Rev. E, **48,29** (1993).

[2] R. Benzi, S. Ciliberto, G. Ruiz Chavarria, C. Baudet Europhysics, **24,** 275 (1993)

[3] A. S. Monin, A. M. Yaglom, Statistical Fluid Mechanics (MIT Press, Cambridge, Ma 1975).

[4] F. Anselmet, Y. Gagne, E. J. Hopfinger, R. A. Antonia, J. Fluid Mech. **140,** 63 (1984).

[5] M. Briscolini, S. Succi, P. Santangelo, R. Benzi submitted to Phys. Rev. Lett.

[6] R. Benzi, L. Tripiccione, F. Massaioli, S. Succi, S. Ciliberto Europhysics Letters, **25**, 341 (1993).

[7] F. Chillá, S. Ciliberto, C. Innocenti, E. Pampaloni Nuovo Cimento **15D**, 1229 (1993).

[8] C. Baudet, S. Ciliberto, Phan Nhan Tien, J. de Physique, **3**, 293 (1993).

Hot Wire Anemometry
(An Overview in Turbulence Research - Present and Future)

A. Tsinober
Department of Fluid Mechanics and Heat Transfer, Faculty of Engineering,
Tel Aviv University, Tel Aviv 69978, Israel

A short account on the method is given with the emphasis on recent developments in attempts to evaluate all components of the velocity derivatives tensor in turbulent air flows (only). The bibliography is compiled in such a way as to provide a guide for those wishing to enter this field and enabling to find an answer practically to any question.

INTRODUCTION

In spite of considerable progress in a variety of modern techniques, such as 3-D particle tracking and holographic methods, hot-wire anemometry (HWA) remains the only technique capable of measuring the smallest and the fastest physically relevant velocity fluctuations in turbulent flows. This is one of the main (but not the only) reasons why HWA remains so popular among researchers and engineers up to now.

The term HWA refers to a variety of techniques, all of which are based on the use of some kind of thermal transducer for the purpose of measuring some property(ies) of fluid flow. Mostly, the property being measured is one or two velocity components. Other properties, such as temperature and composition can be measured, since the transducers are sensitive to heat transfer between the sensor and the environment.

HWA technique is about 100 years old (for an historical introduction see *Comte-Bellot 1976*). During this period a vast literature on the subject has emerged (*Freymuth 1978* has compiled an outstanding bibliography for the period between 1817 and 1978). This consists of at least many hundreds of papers on specific subjects, many review papers and chapters in books, and several books. Only a small part of these are given in the section of references with the emphasis on publications of general nature, reviews, books and papers of special interest from the point of this article. However, it is a must to specially mention the work of *Dryden & Kuethe 1929*, which started the era of *quantitative* measurements of turbulent velocity fluctuations.

A typical modern HWA arrangement consists of a probe/sensor (or several probes), calibration, data acquisition and processing systems. The main part of a HWA probe consists of one or several miniature metallic elements (wires, films), whose *electrical resistance is a function of temperature*. Among several methods of obtaining velocity from the sensor output, the constant temperature method proved to be the most reliable and effective. Its main feature is constant resistance compensating electronic circuitry with a feedback loop, which keeps the temperature of the sensor constant under changing heat transfer conditions due to fluctuating velocity. The compensating voltage or current outputs are 'turned' into velocity via some calibration procedure, which is supposed to produce a one to one relation between the output voltages and velocity components. In case when in the flow exist variations of properties other than velocities, the calibration procedure has to be capable to separate the velocity signal from other influences, the most problematic of which are the variations of temperature. A *really good* calibration of a HWA probe (even with a single wire) is one of the most difficult aspects of HWA, since —citing *Perry 1982*— *the central difficulty with a hot-wire anemometer is that there is no certain way of subjecting the sensor of the instrument to accurately known velocity fluctuations over the desired frequency range of operation.* It is generally accepted that due to very high frequency response the so-called static calibration — when one obtains the necessary relations at 'zero frequency', i.e. in a steady and turbulence free homogeneous flow— is precise enough in most cases (however, see e.g. *Perry 1982, Lebedeva & Dragan 1991*).

There exists a great number of other difficulties in using accurately the HWA technique. These have been carefully discussed in *Perry 1982, Antonia 1993* and references therein. Some of the most important difficulties relevant for this communication will be mentioned in the sequel.

SCOPE AND MOTIVATION

In the following we are concerned with the possibilities of using HWA technique for measurement —or at least evaluation— of the whole tensor of the velocity derivatives along with the three components of the velocity vector. There are several reasons for obtaining *simultaneously* all the velocity components and all the nine velocity derivatives comprising the tensor of velocity derivatives.

Partial information about velocity vector and the tensor of velocity derivatives may be inadequate in several respects (*Tsinober, Kit & Dracos 1991, 1992, Antonia 1993*). For example, it is a common assumption in turbulence research that turbulent dissipation is well represented by a single squared velocity derivative. Except of their mean values other properties of these two quantities are *different even in homogeneous and quasi-isotropic flows.* This problem has been formulated by *Gibson* and *Masiello 1972* (see also *Sreenivasan et al. 1977*), while the first experi-

mental indications of such differences were obtained by *Freytag 1978*.

This points to the necessity of simultaneously measuring all the velocity derivatives allowing to obtain the true dissipation and not its surrogate. An important point is that the full dissipation is a geometrical invariant quantity, which is independent of the system of reference. Such quantities are the most appropriate for describing physical processes. There exists many other invariant quantities of extreme importance in turbulence research, which can be obtained easily from the velocity vector and the tensor of velocity derivatives, e.g. enstrophy (squared vorticity), enstrophy generation, helicity, etc. For example, using such information about enstrophy generation it was possible to confirm by a *direct* experimental observation the prevalence of the vortex stretching process over the vortex compressing in turbulent flows. Another example is that having the information on the whole tensor of velocity derivatives one can evaluate various (statistical) properties of such an invariant quantity as $\Delta \equiv (\partial u_i/\partial x_j)(\partial u_j/\partial x_i)$ and thereby obtain information about the Laplacian of *pressure* (since for incompressible fluids $\nabla^2 p/\rho \equiv \Delta$) and compare its behavior with the results of direct numerical simulations in the case of a Gaussian velocity field, which can be obtained analytically. Properties of Δ are of particular interest also for another two reasons: first $-2\Delta \equiv \omega^2 - \epsilon/\mu$ characterizes the differences between enstrophy and dissipation, and second, Δ represents those terms in the expression for the instantaneous dissipation which make the largest contribution when the purely positive terms are small, since $2\epsilon/\mu \equiv \sum_{i,j}(\partial u_i/\partial x_j)^2 + \Delta$ (*Shtilman et al. 1993b*). Finally, information about *all* the velocity components and their derivatives allows to obtain various invariant geometrical relations of extreme dynamical significance such as alignments between various quantities (velocity - vorticity, vorticity - eigenvectors of the rate of strain tensor, vorticity - vortex stretching vector and several others (see e.g. *Shtilman et al. 1993b, Tsinober et al. 1992, Tsinober 1993* and references therein)).

There still remains the question: why to bother of *measuring* such difficult quantities in the age of super-computers. First, it is possible to perform direct numerical simulations of the Navier-Stokes equations (NSE) at rather low Reynolds numbers in simplest geometries only and the prospects for higher Reynolds numbers are rather pessimistic and not only for the nearest future. Though the scale resolution problem is serious both in laboratory and numerical experiments there is an *essential difference* between the two: inadequate resolution in numerical experiments results in erroneous results, whereas in laboratory/field experiments one has *already* the *true* flow and correct results for the scales resolved even when some range of scales is not resolved. Second, it is still not at all obvious that at *high enough Reynolds numbers* the NSE are describing adequately the turbulent flow (see e.g. *Tsinober 1993* and references therein).

PROBES

Probes for Measurements of all the Three Velocity Components

In order to measure all the three velocity components one needs a probe having at least three wires, not all of them located in one plane. There have been several attempts to build and use triple sensor wires of various configurations (*Buchave 1978, Lakshminarayana 1982, Lekakis et al 1989, Moffat et al. 1978, Paulsen 1989, Sammler et al. 1993* and references therein)[†]. All the three sensor probes perform not satisfactorily when the instantaneous velocity vector forms a small angle with one of the wires of the probe. Therefore, several attempts have been made to use four wire probes and to exploit in some way the redundant information obtained in such a way (*Döbbeling et al. 1990, Pailhas*

[†] Triple-sensor probes are now available from DANTEC and TSI.

& Cousteix 1986, Schön & Müller 1989, Samet & Einav 1987).

Manufacturing of such probes is a pretty delicate matter and requires special equipment and skills. Nevertheless since the commercially available probes are rather expensive (also for repairing) as well as for other reasons such as inconvenience of repairing (they are rather fragile), many prefer to have in house facilities for probe manufacturing and repairing.

It is noteworthy that the motivation for using such probes comes not only from the obvious desire of obtaining all the three velocity components, but also since there are serious indications that e.g. X-probes are unable to capture properly the velocity component normal to the mean flow (see *Antonia 1993* and references therein).

Probes for Evaluation of All the Components of the Tensor of Velocity Derivatives

So far there are known only three attempts to built such probes (*Freytag 1978, Kit et al. 1993, Tsinober et al. 1991, 1992, Vukoslavcević et al. 1991* and *Wallace et al. 1992*). Essentially these probes consist of several arrays, each measuring all the three velocity components in adjacent points in the plane normal to the mean flow (sometimes almost normal to the mean flow). This allows to evaluate the velocity derivatives in this plane taking finite differences between corresponding points, while the derivatives in the direction of the mean flow are obtained via the Taylor hypothesis. Arrays with only three sensors were used by *Balint et al 1991* and *Freytag 1978*. Four-sensor arrays were used by *Kit et al. 1993, Tsinober et al. 1991, 1992* and *Wallace et al. 1992*. An additional feature of probes consisting of five-array probe (*Kit et al. 1993, Tsinober et al. 1991. 1992*) is that *all* first velocity derivatives can be evaluated at *the same point* and that the *second derivatives* of the velocity components can also be evaluated at the same point. Manufacturing of such probes is even more difficult not only due to the very complicated structure of such probes but also due to the necessity to miniaturize each array as much as possible. Without entering into the details (an example of a description of a manufacturing procedure can be found in *Tsinober 1988*) one important aspect should be especially emphasized. One of the ways to somewhat reduce the scale of an array is to use a common prong in each array. However, this leads to electronic cross-talking between the wires in the array which may cause very serious errors. For example in the probe of *Vukoslavcević et al. 1991* the common resistance was about 0.1 Ω which is two orders of magnitude larger than the common resistance allowing to neglect the effects of electronic cross-talking (*Moffat et al. 1978*). According to our checks (*Tsinober et al. 1992*) this *only* can lead to errors in evaluating the velocity derivatives exceeding 40%. Therefore the results obtained with such a probe (and published in several papers by this group) seem to be unreliable.

CALIBRATION

Mostly calibration procedures are based on some universal (or quasi-universal) relations for heat transfer. While this approach seems to be acceptable for velocity measurements— at least for measurements of the streamwise velocity component— (however see *Bradshaw 1971*, p.116), it seems to be not satisfactory in the case of evaluating the velocity derivatives. This is due to a great number of uncertainties about the properties of probe and fluid and to many outside influences to which the probe (and the electronics too) are subjected (for a long list of such influences see *Bradshaw 1971, Lomas 1986, Perry 1982, Sandborn 1972* and numerous references on HWA). Moreover such calibrations employ restrictive assumptions about the symmetry or other geometrical properties of each array as a consequence of using the effective velocity approach, etc. This results in very strict requirements on the geometrical precision of the probe and

its alignment with the flow. Unavoidable imperfections in both lead to *uncontrollable* errors. Finally, rather limited information (i.e. in the plane corresponding to zero pitch and yaw angles) is being used in such calibrations, leading again to uncontrollable errors. Therefore, it is much safer to use an *ad hoc* calibration for *each* experiment with a large amount of 3D calibration data in velocity, pitch and yaw angles with appropriate 3D approximation providing the relations between the sensor outputs and velocity components at each array. Quasi-2D approximations have been employed by *Samet & Einav 1987* and *Döbbeling et al. 1990*. A 3D approximation has been made by *Tsinober et al. 1991, 1992* using Chebychev polynomials. This approximation worked well in the range of $\pm12^{\mathrm{o}}$ only for pitch and yaw angles. Recently *Lemonis & Dracos 1993* have produced a much better approximation on the basis of Bernstein–Bézier polynomials, which works in the range at least $\pm30^{\mathrm{o}}$ for the pitch and yaw angles. Such calibrations require a fully automatic computer controlled calibration unit (*Lemonis & Dracos 1993, Tsinober et al. 1992*). Since the calibration is made in a homogeneous flow, errors arise due to velocity gradients on the scale of the whole probe (typically about 3 *mm*) and on the scale of individual arrays. Errors arising in such a way are particularly serious in regions with strong gradients of velocity. Though some authors claim to be able to take into account the velocity gradients on the scale of the probe, this problem remains open, the best (but extremely difficult) solution of which is considerable reduction of scale of the probe (see below).

ELECTRONICS, DATA ACQUISITION SYSTEMS AND TESTS

At present the electronics and data acquisition systems are rather standard and of high quality. The main problems are with probes and calibration systems (i.e. both calibration hardware and software).

An aspect of utmost importance in using HWA generally and especially with multi-sensor/multi-array probes for measurements velocity derivatives is making a set of tests of performance of the whole system and its separate elements, especially probes and calibration. Most of these are described in the cited literature. This set of tests was enriched recently by comparisons of the laboratory results with those obtained via direct numerical simulations (*Antonia 1993, Tsinober 1993*), which is possible for rather small Reynolds numbers only. However, these comparisons have to be taken with care for several reasons. One such reason is that, mostly, DNS are performed employing periodic boundary conditions at least at some of the boundaries of the flow domain. Therefore, the correlation coefficient between two values of *any* quantity at the opposite such boundaries (i.e. the points separated by a *maximal distance* in the flow domain) will be precisely equal to unity and close to unity for the points in the proximity of such boundaries, whereas in any *real* flow the correlation coefficient becomes very small for points separated by a distance of the order of the integral scale of turbulent flow.

PRESENT VERSUS FUTURE

It is our opinion and experience that the present status of the multisensor HWA for measurements of velocity gradients is such that one can obtain these and related quantities with errors of about 30% in case of probes *without common prongs*. Therefore, only strong qualitative effects can be studied reliably (for example, see *Tsinober et al. 1992*), except, perhaps, for some scaling properties. For example, one cannot seriously check the small scale (an)isotropy in turbulent flows using presently existing techniques *even with probes without common prongs*, though such an intention has been announced by *Wallace et al. 1992*.

One of the problems is a reliable 3D calibration procedure. There are two ways of considerably improving the existing calibration systems. One is to use such a system *in situ continuously* using the 'closest' calibration files to the experimental

run right before and after *with pitching and yawing of the calibration flow and not the probe*. The other is to *continuously* compare the response of the hot wire probe (which is a fast one) to that of a slower one whose calibration does not vary such as a sonic anemometer in a manner similar to that discussed by *Mestayer et al. 1990* and *Onsley 1989*. Both methods are appropriate in field experiments and are presently in the process of development: the first one in our Laboratory and the second one in the Laboratory for Aero/Hydrodynamics in Delft (*van Dijk et al. 1993*). Along with the approximation developed by *Lemonis & Dracos 1993* such calibration procedures are expected to be several times more precise.

Another problem is the spatial resolution multi-array/multi-sensor probes. As mentioned the typical scale of such a probe is about 3 *mm*. This means that in order to resolve all the scales down to the Kolmogorov one —which is typically about several tenths of *mm*— one has to reduce the scale of the multi-array/multi-sensor probe by *an order of magnitude*. According to our experience it is possible to manufacture a probe with the overall scale of about 1 *mm* using very fine wires (0.6 micrometer thick —such probes consisting of single wire and X-probes have been made by *Ligrani et al. 1989*). Even this requires considerable increase of investment into appropriate equipment. Smaller than 1 *mm* multi-array/multi-sensor probes *can be manufactured on a totally different principle* and correspondingly require at least an order of magnitude larger investment.

The HWA technique for measurements, especially of velocity derivatives *still has more of the caprice of an art than the complete reliability of a routine laboratory procedure* (*Kovasznay 1959*) and *it is this 'caprice of the art' which leads many people not only worry about the calibration of the instrument but also the person carrying out the measurements* (*Perry 1982*).

Nevertheless, the HWA remains the only technique which *is capable* to resolve the smallest scales in turbulent flows with *large* Reynolds numbers. Apart of the reasons given above and general *fundamental importance of high quality of such measurements* there is an urgent need due to massive use of large eddy simulations for computations of large Reynolds turbulent flows, the very basis of which requires *reliable* information on the properties of the small scales in such flows. There is no way other than experiment to obtain such information which so far does not exist. Therefore, no investment seems to be exaggerated for such an endeavor.

Bibliography

Books

[1] C.G. Lomas 1986 *Fundamentals of Hot Wire Anemometry*, 211 pp. Cambr. Univ. Press.

[2] A. Perry 1982 *Hot-Wire Anemometry*, Clarendon, Oxford. V.A. Sandborn 1972 *Resistance Temperature Transducers*, 346 pp. Metrology Press, Fort Collons, Colorado.

[3] A.V. Smolyakov & V.M Tkachenko 1983 *The Measurement of Turbulent Fluctuations - An Introduction to Hot-Wire Anemometry and Related Transducers*, 298 pp. Springer.

Review papers and chapters in books

[4] P. Bradshaw 1971 *An Introduction to Turbulence and its Measurement*, Ch. 5, The Hot-Wire Anemometer, pp. 106 - 133, Pergamon.

[5] G. Compte-Bellot 1976 Hot-Wire Anemometry, *Ann. Rev. Fluid Mech.*, **8**, pp.209 - 231.

[6] S. Corrsin 1963 Turbulence: Experimental Methods, in *Handbuch der Physik*, **VIII/2**, ed. S. Flügge, pp. 524 - 590, Springer.

[7] J.F. Foss and J.M. Wallace 1989 The Measurement of Vorticity in Transitional and Fully Developed Turbulent Flows. *Advances in Fluid Mechanics Measurements*, ed. M. Gad-el Hak, *Lect. Notes Engn.*, **45**, 263 - 321.

[8] P. Freymuth 1978 A Bibliography on Thermal Anemometry, *TSI Quarterly*, Iss. 4.

[9] J.O. Hinze 1975 *Turbulence*, Ch. 2 , Principles of Methods and Techniques in the Measurement of Turbulent Flows, pp. 83 - 174, McGraw-Hill.

[10] J.- D. Vagt 1979 Hot-Wire Probes in Low Speed Flow, *Prog. Aerospace Sci.*, **18**, 271 - 323.

[11] L.S.G. Kovasznay 1959 Turbulence measurements, *Appl. Mech. Rev.*, **12**, 375 - 379.

Proceedings of Specialized Meetings

[12] Proceedings of the Dynamic Flow Conference 1978 on *Dynamic Measurements in Unsteady Flows*, ed. L.S.G. Kovasznay, Skovlunde, Denmark.

[13] Thermal Anemometry 1993 *The Fluids Engn. Conf., Washington, D.C. 1993*, ASME, *FED* - **Vol. 167**.

Specific References

[14] R.A. Antonia 1993 Direct Numerical Simulations and Hot wire Experiments: A possible Way Ahead? *New Approaches and Concepts in Turbulence-Proceedings of the Monte Verità Colloquium*, eds. T. Dracos and A. Tsinober, 349 - 365.

[15] J.-L. Balint, J.M. Wallace and P. Vukoslavcević 1991 The Velocity and Vorticity Vector Fields of a Turbulent Boundary Layer. Part 2. Statistical Properties, *J. Fluid Mech.*, **228**, 53 - 86.

[16] P. Buchave 1978 Transducer Techniques, *Proceedings of the Dynamic Flow Conference 1978 - Dynamic Measurements in Unsteady Flows*, ed. L.S.G. Kovasznay, pp. 427 - 463, Skovlunde, Denmark.

[17] A. van Dijk, A. Tsinober, F. Nieuwstadt, G. Lemonis and T. Dracos 1993 In Situ Calibration of a Multi-Hotwire Probe Using a Sonic Anemometer, to be presented at the *Fifth European Turbulence Conference*, Siena, July 5 -8, 1994.

[18] K. Döbbeling, B. Lenze and W. Leukel 1990 Computer-Aided Calibration and Measurements with a Quadruple Hotwire Probe, *Exp. Fluids*, **8**, 257 - 262.

[19] H.L. Dryden and A.M. Kuethe 1929 The Measurement of Fluctuations of Air Speed by the Hot-Wire Anemometer, *NACA Rep.*, **320**.

[20] C. Freytag 1978 Statistical Properties of Energy Dissipation, *Boundary-Layer Meteorology*, **14**, 183 - 198.

[21] C.H. Gibson and P.J. Masiello, 1972 Observations of the variability of dissipation rates of turbulent velocity and temperature fields, *Statistical Models and Turbulence*, ed. M. Rosenblatt and C. Van Atta, Springer, 431 - 442.

[22] E. Kit, A. Tsinober and T. Dracos 1993 Velocity gradients in a Turbulent Jet Flow, *Appl. Sci. Res.*, **51**, 185 - 190.

[23] B. Lakshminarayana 1982, Three-Sensor Hot-Wire/Film Technique for Tree Dimensional Mean and Turbulence Flow Field Measurement, *TSI Quarterly*, **8**, No.1. 3- 13; see also B. Lakshminarayana and R. Davino 1988 Sensitivity of Three Sensor Hot Wire Probe to Yaw and Pitch Angle Variation, *Trans. ASME - J. Fluid Engn.*, **120**, 120 - 122.

[24] I.C. Lekakis, R.J. Adrian and B.J. Jones 1989 Measurement of Velocity Vectors with Orthogonal and Non-Orthogonal Triple-Sensor Probes, *Exp. Fluids*, **7**, 228 - 240.

[25] G. Lemonis and T. Dracos 1993 A New Calibration and Data Reduction Method for Turbulence Measurement by Multi-hotwire Probes, *Experiments in Fluids*, submitted.

[26] P.M. Ligrani, R.V. Vestpal and F.R. Lemos 1989 Fabrication and Testing of Subminiature Multi-Sensor Hot-Wire Probes, *J. Phys. E: Sci. Instrum.*, **22**, 262 - 268.

[27] R.J. Moffat, S. Yavuzkurt and M.E. Crawford 1978 Real-Time Measurements of Turbulence Quantities With a Triple Hot-Wire System *Proceedings of the Dynamic Flow Conference 1978 - Dynamic Measurements in Unsteady Flows*, ed. L.S.G. Kovasznay, pp. 1013 - 1035, Skovlunde, Denmark.

[28] P.G. Mestayer, S.E. Larsen, C.W. Fairall and J.B. Edson 1990 Turbulence Sensor Dynamic Calibration Using Real-Time Spectral Computations, *J. Atm. Ocean. Techn.*, **7**, 841 - 850.

[29] S.P. Oncley 1989 Flux Parameterization Techniques in the Atmospheric Surface Layer, Ph.D. Thesis, University of California - Irvine.

[30] G. Pailhas and J. Cousteix 1986 Method for Analyzing Four-Hot-Wire Probe Measurements, *Rech. Aérosp*, **2**, 79 - 86.

[31] L. Paulsen 1983 Triple Hot-Wire Technique for Simultaneous Measurements of Instantaneous Velocity Components in Turbulent Flows, *J. Phys. E: Sci. Instrum.*, **16**, 554 - 562.

[32] M. Samet and S. Einav, 1987 A Hot-Wire Technique for Simultaneous Measurement of Instantaneous Velocities in 3D Flows, *J. Phys. E: Sci. Instrum.*, **20**, 683 - 690.

[33] B. Sammler, R. Eschenhagen, K. Gaichen, H. Kitzing and G. Seifert 1993 Multi-Hot-Wire-Probes for Investigations of Turbulence, *Thermal Anemometry 1993 - The Fluids Engn. Conf., Washington, D.C. 1993*, ASME, *FED -* **Vol. 167**, 231 -240.

[34] L. Shtilman, M. Spector, A. Tsinober and H. Vaisburd 1993b A study of properties of cross-product of velocity derivatives, *in preparation*.

[35] T. Schön and U. R. Müller 1989 A New Hot-wire Technique for Measuring the Instantaneous Velocity Vector in Highly Turbulent Flow, *Advances in Turbulence* **2**, eds. H.- H. Fernholz and H.E. Fiedler, pp. 292 - 297.

[36] K.R. Sreenivasan, R.A. Antonia and H.Q. Danh, 1977 Temperature dissipation fluctuations in a turbulent boundary layer, *Phys. Fluids*, **20**, 1238 - 1245.

[37] A. Tsinober 1988 Multi-Hot-Wire Probe Production for Measurements of All Nine Velocity Derivatives, *Int. Rep., Fac. Engn., Tel Aviv Univ.* - An updated version is available upon request.

[38] A. Tsinober, 1993 Some properties of velocity derivatives in turbulent flows as obtained from experiments: laboratory and numerical, in *Turbulence in Spatially Extended Systems*, eds. R. Benzi, C. Basdevant and S. Ciliberto, Nova Science, (in press).

[39] A. Tsinober, E. Kit and T. Dracos 1991 Measuring Invariant (Frame Independent) Quantities Composed of Velocity Derivatives in Turbulent Flows. *Advances in Turbulence*, **3**, ed. A.V. Johansson & P.H. Alfredson, pp. 514 - 523, Springer.

[40] A. Tsinober, E. Kit and T. Dracos 1992 Experimental Investigation of the Field of Velocity Gradients in Turbulent Flows, *J. Fluid Mech.*, **242**, 169 - 192.

[41] P. Vukoslavcević, J.M. Wallace an J.- L. Balint 1991 Vorticity Vector Fields of a Turbulent Boundary Layer. Part 1. Simultaneous Measurement by Hot-Wire Anemometry, *J. Fluid Mech.*, **228**, 25 - 51.

[42] J.M. Wallace, L. Ong and J.-L. Balint 1992 An investigation of Small Scales of Turbulence in a Boundary Layer at High Reynolds Numbers, *Annual Research Briefs*, pp.263 - 268, Center for Turbulence Research, Stanford,

INTERMITTENCY
(Random Cascade Models, Multifractality and large Deviations)

U. FRISCH

C.N.R.S., Observatoire de Nice,

B.P. 229, F-06304 Nice cedex 4, France

INTRODUCTION

This is not meant to be a self-contained exposition on intermittency and multifractality in fully developed turbulence. For such matters, the reader is referred to Mandelbrot (1974)[15], Monin and Yaglom (1975)[21], Frisch, Sulem and Nelkin (1978)[11], Meneveau and Sreenivasan (1991)[20], Frisch (1991)[9], Aurell *et al.* (1992)[2] and references therein. A detailed exposition will be found in the forthcoming book by Frisch (1995)[10]. Let us just make some brief preliminary remarks.

Models having "intermittent" corrections to the Kolmogorov 1941 scaling were introduced by the Russian school (Kolmogorov, Obukhov, Yaglom, Novikov in collaboration with Stewart) in the sixties. They were meant to give a more accurate description of the small scales, which display significant spatio-temporal fluctuations of the energy dissipation. Such work used mostly multiplicative random cascade models of the kind discussed in the next section. Mandelbrot (1974)[15] showed that these models give rise to an energy dissipation, the support of which is concentrated on a fractal set. Parisi and Frisch (1985)[25] suggested the modern concept of "multifractal" in which there is a whole range of singularity exponents h of the velocity field, each living on a different fractal set, having a dimension $D(h)$. A similar idea was used by Benzi *et al.* (1984)[3] and by Halsey *et al.* (1986)[12] to describe the structure of invariant measures of attractors in the phase space of dynamical systems. Meneveau and Sreenivasan

(1987)[19] used this formalism to describe the multifractal properties of the (space-averaged) energy dissipation with a dimension $F(\alpha)$ depending on the singularity exponent α of the dissipation. Space-averaged fluctuations of the energy dissipation and fluctuation in velocity increments can be related in a somewhat heuristic way, using a suggestion of Kolmogorov (1962)[13]. For this, he was essentially relying on his 1941 result stating that velocity increments δv_ℓ over a distance ℓ scale as $(\epsilon\ell)^{1/3}$, where ϵ is the mean energy dissipation. Kolmogorov proposed to use the same formula with ϵ_ℓ instead of ϵ. Here, ϵ_ℓ is the energy dissipation space-averaged over a ball (or cube) of radius (or side) ℓ. Specifically, he assumed that the scaling exponent ζ_p for the moment of order p of velocity increments is the same as the scaling exponent for the moment of order p of $(\epsilon_\ell\ell)^{1/3}$. In (singularity) multifractal language this can be formulated by saying that to any singularity of exponent α of $\epsilon_\ell\ell$, there is an associated singularity of exponent $h = \alpha/3$ for the velocity on the same set. Thus, one has the following "dictionary" between the two multifractal formalisms:

$$
\begin{aligned}
h &= \frac{\alpha}{3}, \\
D(h) &= F(\alpha), \\
\zeta_p &= \frac{p}{3} + \tau_{p/3},
\end{aligned}
\tag{1}
$$

where it is assumed that the moment of order q of $\epsilon_\ell\ell$ scales as ℓ^{τ_q}.

The connection between random cascade models and the modern multifractal formalism follows from the theory of large deviations. This theory goes back to the late thirties and its relevance for turbulence has been realized only rather recently (Mandelbrot 1989[16], 1991[18]; Oono 1989[24]; Collet and Koukiou 1992[5]). It will be presented in the last section.

RANDOM CASCADE MODELS

We begin with a cube of side ℓ_0 in which we take the dissipation to be uniform equal to ϵ, a nonrandom positive quantity. This defines the generation $n = 0$ of the model. To obtain generation $n = 1$, we subdivide the cube into eight equal cubes of side $\ell_1 = \ell_0/2$. (The factor of two is assumed only for simplicity.) In each smaller cube we multiply the dissipation by independent realizations of a random variable W, which is subject to the following constraints:

$$W \geq 0, \quad \langle W \rangle = 1,$$
$$\langle W^q \rangle < \infty, \ \forall q > 0. \tag{2}$$

At the nth generation, there are 2^n cubes of side

$$\ell = \ell_0 2^{-n}. \tag{3}$$

In each of them the dissipation is uniform and assumes a value of the form

$$\epsilon_\ell = \epsilon W_1 W_2 \ldots W_n, \tag{4}$$

where the W_i's are independently and identically distributed. The process is repeated indefinitely. In the limit, we obtain for the dissipation a highly singular positive random measure ϵ_∞. The quantity ϵ_ℓ will be called the ℓ-averaged dissipation. By (2), the (ensemble) average of any of the ϵ_ℓ's is still equal to ϵ but, as we shall see, there are large fluctuations.

It is a straightforward matter to calculate moments of ϵ_ℓ. From (3) and (4), we obtain for any positive integer q

$$\langle \epsilon_\ell^q \rangle = \epsilon^q \left(\frac{\ell}{\ell_0} \right)^{\tau_q}, \tag{5}$$

where

$$\tau_q = -\log_2 \langle W^q \rangle. \tag{6}$$

From the bridging relation (1) we obtain the following expression for the exponents of structure functions, i.e. moments of velocity increments:

$$\zeta_p = \frac{p}{3} - \log_2 \langle W^{p/3} \rangle. \tag{7}$$

A particularly simple result is obtained with the *black and white* choice of Novikov and Stewart (1964)[22], a term suggested by Mandelbrot (1974)[15]. Here, only two values are permitted for the random variable W,

$$W = \begin{cases} 1/\beta & \text{with probability } \beta \\ 0 & \text{with probability } 1 - \beta \end{cases} \tag{8}$$

From (6) and (7), we obtain for the Novikov-Stewart model

$$\tau_q = -(1 - q) \log_2 \beta \tag{9}$$
$$\zeta_p = \frac{p}{3} - \left(1 - \frac{p}{3} \right) \log_2 \beta, \tag{10}$$

which is exactly the result for the (unifractal) β-model (Frisch, Sulem and Nelkin 1978[11]).

Benzi *et al.* (1984)[3] have proposed a *random β-model* in which the factor β is randomly and independently selected at each step of the cascade, with the same law. Just as the ordinary β-model is equivalent to the Novikov-Stewart model, as far as scaling is concerned, making an *arbitrary* random choice of β is equivalent to the general random cascade model. Benzi *et al.* (1984)[3] also proposed a restricted choice with a single free parameter x between zero and one, for which the p.d.f. of β is $P(\beta) = x\delta(\beta - 0.5) + (1 - x)\delta(\beta - 1)$. A good fit to the experimental data of Anselmet *et al.* (1984)[1] is obtained for $x = 0.125$.

The graph of τ_q is a straight line in the Novikov-Stewart model; however, as soon as W takes more than one non-vanishing value, (6) produces a non-trivial concave graph, a consequence of the Schwartz inequality (see Section V.8 of Feller, 1966[8]). Hence, the random cascade models have usually *multifractal* scaling properties. In order to gain a deeper understanding of why such a scaling holds, we present now, in a simplified form, the theory of large deviations.

LARGE DEVIATIONS AND MULTIFRACTALITY

Large deviations theory is concerned with the very low probability events where the sums or integrals of random variables or functions differ from their average values by an amount much larger that the standard deviation. The theory originated with the work of Cramér (1938)[6]. For a general exposition we refer to Varadhan (1984)[26] and Ellis (1985)[7]. For the application to random cascade models, we shall need only the case of independent variables, which is now briefly reviewed and we shall here follow the exposition of Lanford (1973)[14].

Let m_i ($i = 1, 2, \ldots$) be identically distributed and independent random variables.[†] We know that under suitable conditions (Feller 1966[8]),

$$S_n = \frac{1}{n} \sum_{i=1}^{n} m_i \to \langle m \rangle \quad for \ n \to \infty,$$

$$(11)$$

the convergence being almost sure (law of large numbers). The theory of large deviations gives estimates of the probability that the partial averages S_n are, for large n, close to an arbitrary value x, which may differ from $\langle m \rangle$ by $O(1)$.

Let

$$h(n; a, b) = \ln \mathrm{Prob}\{a < S_n < b\}. \quad (12)$$

The function $h(n; a, b)$ is non-positive and *superadditive*, i.e.,

$$h(n + n'; a, b) \geq h(n; a, b) + h(n'; a, b). \quad (13)$$

This follows immediately from the observation that, by the assumed independence of the m_i's:

$$\mathrm{Prob}\{a < \frac{1}{n + n'} \sum_{1}^{n+n'} m_i < b\} \geq$$

$$\mathrm{Prob}\{a < \frac{1}{n} \sum_{1}^{n} m_i < b\} \times$$

$$\mathrm{Prob}\{a < \frac{1}{n'} \sum_{n+1}^{n+n'} m_i < b\} \quad (14)$$

A simple consequence of superadditivity is that the limit

$$\begin{aligned} s(a, b) &= \lim_{n \to \infty} \frac{h(n; a, b)}{n} \\ &= \sup_{n} \frac{h(n; a, b)}{n} \end{aligned} \quad (15)$$

exists. [Hint: Pick two integers n and n_0, perform the Euclidian division $n = qn_0 + r$, apply superadditivity and then let successively n and n_0 tend to infinity.] The value $-\infty$ is allowed for the limit. We then define

$$s(x) = \sup_{a,b} s(a, b), \quad for \ a < x < b. \quad (16)$$

The function $s(x)$ is negative or zero (by construction) and is easily shown to be concave.

The large deviations theorem, loosely stated, is that for large n (if you are a mathematician, skip the next line)

$$\mathrm{Prob}\{\frac{1}{n} \sum_{i=1}^{n} m_i \approx x\} \sim \exp[s(x)n]. \quad (17)$$

In statistical mechanics, large deviations theory is used to prove rigorous results about the approach to the thermodynamic limit. In this context the function $s(x)$ can be identified with the *entropy*; in the literature on large deviations, the negative of $s(x)$ is often called the 'rate function'. Mandelbrot (1991)[18] has proposed to call $s(x)$ the *Cramér function*. This seems fully justified and we shall adopt this terminology here.

A simple example of large deviations (also taken from Lanford, 1973[14]) is the coin-tossing problem. The variable m takes only two values, say 0 and 1, with equal probabilities. The probability of S_n can be explicitly calculated from the binomial formula. Use of Stirling's formula on

---

[†]In the application to random cascade models, $m = - \log_2 W$.

41

factorials, gives then the following explicit expression:

$$s(x) = \begin{cases} f(x) & \text{for } 0 \le x \le 1 \\ -\infty & \text{otherwise} \end{cases} \qquad (18)$$

(Here,
$f(x) = -x \ln x - (1-x) \ln(1-x) - \ln 2$.)
Note that the function $s(x)$ defined by (18) behaves parabolically near its maximum $x = 1/2$, $s = 0$. This is quite general and equivalent to the statement that deviations which are $O(n^{-1/2})$ have a Gaussian distribution. Larger deviations, which are $O(1)$, cannot be correctly described as Gaussian.

In the general case, the Cramér function $s(x)$ can be expressed in terms of the logarithm of the characteristic function of the random variable m. Let us assume that the p.d.f. of the random variable m decreases faster than exponentially at large arguments; this ensures the existence of

$$Z(\beta) = \langle e^{-\beta m} \rangle, \qquad (19)$$

for any real β. The latter is the characteristic function for an imaginary argument $z = i\beta$.

The statement is that the functions $s(x)$ and $\ln Z(\beta)$ are Legendre transforms of each other:

$$\begin{aligned} \ln Z(\beta) &= \sup_x (s(x) - \beta x), \qquad (20) \\ s(x) &= \inf_\beta (\ln Z(\beta) + \beta x). \quad (21) \end{aligned}$$

A rigorous but still elementary proof of (20) and (21) may be found in Lanford (1973)[14]. Here, we give a simplified proof based on (17). We observe that

$$Z^n(\beta) = \langle e^{-\beta(m_1 + \ldots + m_n)} \rangle. \qquad (22)$$

By the large deviations theorem, for large n, the sum $m_1 + \ldots + m_n$ is close to nx with a probability $\sim \exp[s(x)n]$. Hence, the contribution to $Z^n(\beta)$ coming from sums close to nx is $\sim \exp[n(-\beta x + s(x))]$. When integrating over all possible x's the dominant contribution will come from that x which maximizes $-\beta x + s(x)$. Thus,

$$Z^n(\beta) \sim \exp\left(n \sup_x (s(x) - \beta x) \right). \qquad (23)$$

Taking logarithms, we obtain (20). This concludes our digression on large deviations.

We return to the random cascade model of the previous section. The dissipation measure ϵ_∞, which is the limit obtained by the construction defined at the beginning of this Section, is *multifractal*. Indeed, setting

$$W_i = 2^{-m_i}, \qquad (24)$$

we obtain from (3), (4) and the large deviations theorem

$$\begin{aligned} \frac{\epsilon_\ell}{\epsilon} &= 2^{-(m_1 + \ldots + m_n)} \\ &\approx 2^{-nx} \approx \left(\frac{\ell}{\ell_0} \right)^x. \qquad (25) \end{aligned}$$

This holds with a probability

$$p_\ell \sim e^{ns(x)} = \left(\frac{\ell}{\ell_0} \right)^{-s(x)/\ln 2}. \qquad (26)$$

In other words, in the nesting construction of cubes, the fraction of the space such that the dissipation behaves as ℓ^x rarefies as $\ell^{-s(x)/\ln 2}$. In the multifractal language of Meneveau and Sreenivasan (1991)[20], (25) and (26) imply that the singularity exponent is $\alpha = x + 1$ and that the codimension $3 - F(\alpha)$ of the singularities with exponent α is $-s(x)/\ln 2$. Hence,

$$\begin{aligned} F(\alpha) &= 3 + \frac{s(\alpha - 1)}{\ln 2}, \\ f(\alpha) &= 1 + \frac{s(\alpha - 1)}{\ln 2}. \qquad (27) \end{aligned}$$

The function $f(\alpha) = F(\alpha) - 2$ is more appropriate when one-dimensional measurements are made, as is usual with a probe. The latter can be viewed as recording the velocity along a one-dimensional cut of the flow (in the reference frame of the mean flow).
Hence, the $f(\alpha)$ function is essentially the Cramér function and should be called so.

We observe that the β-model of Frisch, Sulem and Nelkin (1978)[11] is obtained when the random variable W takes only two values: $1/\beta$ (with probability β) and zero (with probability $1 - \beta$). By (24), the corresponding values for m are $\ln_2 \beta$ and $+\infty$. This gives a somewhat pathological character to the large deviations result. Indeed, the normalized sums S_n, defined in (11), also take only two values: $\ln_2 \beta$ (with probability β^n) and $+\infty$ (with probability $1 - \beta^n$). Hence, there is a *single* value for the 'deviation' x and a *single* value for the dimension D, given by $3 - D = -\ln_2 \beta$.

The dimension $F(\alpha)$ given by (27) may well become *negative* for some range of α's. This means that the nested set of cubes on which such an α holds, does not terminate, because it rarefies too quickly with the generation index n. Still, $F(\alpha)$ (or, equivalently, $s(x)$) is meaningful as a quantity controlling the probability of such rarefactions (through (17)). For the concept of negative dimension, see Mandelbrot (1990, 1991)[17, 18].

Large deviations theory, was here presented in its simplest form, for independent random variables. Like the law of large numbers and the central limit theorem, large deviations theory has extensions to random variables with correlations, when the latter decrease sufficiently fast. Large deviations can also occur in deterministic chaos, but need not. Indeed, Biferale *et al.* (1994)[4] have constructed deterministic cascade models in which the factors W_n of (4) are obtained by iterating a deterministic chaotic map. These models have chaotic fluctuations, but do not deviate from the K41 scaling because they have no large deviations.

Bibliography

[1] Anselmet, F., Gagne, Y. Hopfinger, E.J. and Antonia, R.A. (1984). High-order velocity structure functions in turbulent shear flow, *J. Fluid Mech.* **140**, 63–89.

[2] Aurell, E., Frisch, U., Lutsko, J. and Vergassola, M. (1992). On the multifractal properties of the energy dissipation derived from turbulence data, *J. Fluid Mech.* **238**, 467–486.

[3] Benzi, R, Paladin, G., Parisi, G. and Vulpiani, A. (1984). On the multifractal nature of fully developed turbulence and chaotic systems, *J. Phys.* **A17**. 3521-3531.

[4] Biferale, L., Blank, M. and Frisch. U. (1993). Chaotic cascades with Kolmogorov 1941 scaling, *J. Stat. Phys.*, **75**, 781-795.

[5] Collet P. and Koukiou, F. (1992). Large deviations for multiplicative chaos, *Commun. Math. Phys.* **47**, 329.

[6] Cramér, H. (1938). *Sur un nouveau théorème-limite de la théorie des probabilités*, vol. 736, in "Actualités scientifiques et industrielles", (Hermann, Paris).

[7] Ellis, R.S. (1985). *Entropy, Large Deviations and Statistical Mechanics*, (Springer).

[8] Feller, W. (1966). *An Introduction to Probability Theory and its Applications*, vol. 2, (Wiley, New York).

[9] Frisch, U. (1991). From global scaling, à la Kolmogorov, to local multifractal scaling in fully developed turbulence, in *Kolmogorov's ideas 50 years on*, eds. Hunt, J.C.R., Phillips, O.M. and Williams, D., *Proc. R. Soc. London* vol. **434**, 89–99.

[10] Frisch, U. (1995). *Turbulence: The legacy of A.N. Kolmogorov*, (Cambridge University Press). *(To appear)*.

[11] Frisch, U., Sulem, P.L. and Nelkin, M. (1978). A simple dynamical model of intermittent fully developed turbulence, *J. Fluid Mech.* **87**, 719–736.

[12] Halsey, T.C, Jensen, M.H., Kadanoff, L.P., Procaccia, I. and Shraiman, B.I. (1986). Fractal measures and their singularities: the characterization of strange sets, *Phys. Rev. A* **33**, 1141-1151.

[13] Kolmogorov, A.N. (1962). A refinement of previous hypotheses concerning the local structure of turbulence in a viscous incompressible fluid at high Reynolds number, *J. Fluid Mech.* **13**, 82–85.

[14] Lanford, O.E. (1973). Entropy and equilibrium states in classical mechanics, in *Statistical Mechanics and Mathematical Problems*, ed. Lenard, A., (Springer) vol. **20**, 1–113.

[15] Mandelbrot, B. (1974). Intermittent turbulence in self-similar cascades: divergence of high moments and dimension of the carrier, *J. Fluid Mech.* **62**, 331–358.

[16] Mandelbrot, B. (1989). Multifractal measures, especially for the geophysicist, *Pure Appl. Geophys.* **131**, 5–42.

[17] Mandelbrot, B. (1990). Negative fractal dimensions and multifractals, *Physica A* **163**, 306–315.

[18] Mandelbrot, B. (1991). Random multifractals: negative dimensions and the resulting limitations of the thermodynamic formalism, *Proc. R. Soc. Lond. A* **434**, 79–88.

[19] Meneveau, C.M. and Sreenivasan, K.R. (1987). The multifractal spectrum of the dissipation field in turbulent flows, *Nucl. Phys. B Proc. Suppl.* **2**, 49-76.

[20] Meneveau, C.M. and Sreenivasan, K.R. (1991). The multifractal nature of turbulent energy dissipation, *J. Fluid Mech.* **224**, 429-484.

[21] Monin, A.S., and Yaglom, A.M. (1975). *Statistical Fluid Mechanics*, vol. 2, ed. Lumley, J., (M.I.T. Press).

[22] Novikov, E.A. and Stewart, R.W. (1964). The intermittency of turbulence and the spectrum of energy dissipation, *Izv, Akad. Nauk SSSR, ser. geoffiz.*, 408–413.

[23] Obukhov, A.M. (1962). Some specific features of atmospheric turbulence, *J. Fluid Mech.* **13**, 77–81.

[24] Oono, Y. (1989). Large deviations and statistical physics, *Progr. Theor. Phys. Suppl.* **99**, 165–205.

[25] Parisi, G. and Frisch, U. (1985). On the singularity structure of fully developed turbulence, in *Turbulence and Predictability in Geophysical Fluid Dynamics, Proceed. Intern. School of Physics "E. Fermi", 1983, Varenna, Italy*, eds. Ghil, M., Benzi, R. and Parisi, G., (North–Holland), 84–87.

[26] Varadhan, S.R.S (1984). *Large Deviations and Applications*, (SIAM).

NUMERICAL SIMULATIONS
(Direct)

M.E. BRACHET

Laboratoire de Physique Statistique,
Ecole Normale Supérieure,
24, rue Lhomond, F-75231 Paris cedex 05, France

Successful modeling of physical phenomena can be formulated as coherence between experiments, theory and computations. Theory provides the conceptual framework and the basic laws. Experiment is concerned with the behavior of real systems. Computations link theory with experiments. In recent years, computational power has increased so greatly that (both theoretical and experimental) research on complex systems, described by simple laws, has relied extensively on insights obtained through direct numerical simulations (DNS).

Such is the case in turbulence research: a turbulent flow does indeed follow simple laws. These laws, for constant density and small Mach number flows, are a system of PDEs called the incompressible Navier Stokes (NS) equations:

$$\partial_t v + (v \cdot \nabla)v = -\nabla p + \nu \nabla^2 v \quad (1)$$
$$\nabla \cdot v = 0. \quad (2)$$

In a wall-bounded domain, the NS equations must be supplemented with no-slip boundary conditions. In general, some kind of external driving is also needed in order to obtain a statistically steady state. When the Reynolds number $Re = UL/\nu$ (where U and L are typical velocity and length scales of the flow, and ν is the kinematic viscosity) is large enough, the system displays turbulent behavior: it is in a time-dependent highly nonlinear regime. With periodic boundary conditions, the NS equations can be solved using spectral methods, with great precision and numerical efficiency (the precision of spec-

tral methods is exponential in the resolution, rather than algebraic for e.g. finite differences methods)[1]. The basic idea is to expand the fields in (truncated) Fourier series, and to compute derivatives in Fourier space and nonlinear terms in physical space. Fast Fourier Transforms (FFT) are used to go back and forth from spectral to physical space in $O(n^D \log(n))$ operations, where D is the space dimension. These methods are quite easy to implement. As a practical illustration, we give here a toy 2D NS pseudo-spectral solver, written in Mathematica[2]. It uses the (Ψ, ω) form of the nonlinear terms, and time-stepping is carried out by the first-order Euler implicit scheme: $u_{n+1} = (1 + \nu k^2 \Delta t)^{-1}(u_n + \Delta t N(u_n))$.

The solver consists of 2 instructions to define the FFTs with proper normalization, and a kernel of 9 instructions (List. 1).

Finally, 7 more instructions are needed to initialize the fields, run the simulation and visualize the result (List. 2).

Running this simple solver produces the visualization shown in Fig. 1.

State-of-the-art spectral solvers are very much like that above. They typically use a second-order time-stepping scheme, such as Leapfrog-Crank-Nicolson: $u_{n+1} = (1 + 2\nu k^2 \Delta t)^{-1}(u_{n-1} + 2\Delta t N(u_n))$, real-to-complex-conjugate FFTs that run faster than general complex FFTs and are de-aliased[1]. As most of the computational time is spent in the FFTs, their number is reduced to a minimum. Codes are generally written in FORTRAN or C,

```
(*Viscosity, time step and resolution*)
nu=.02;dt=.01;n=64

(*FFTs*)
PhysicalToSpectral[l_]:=InverseFourier[l]/n
SpectralToPhysical[l_]:=n Fourier[l]

(*Wavenumbers*)
Kwave=Table[I (Mod[i-1+n/2,n]-n/2),{i, n}];
KwaveSquare=Kwave^2;
Kwave[[n/2+1]]=0;

(*Derivatives*)
xDerivate[l_]:=Table[l[[i,j]] * Kwave[[i]],{i,n},{j,n}]
yDerivate[l_]:=Table[l[[i,j]] * Kwave[[j]],{i,n},{j,n}]

Laplacien[l_]:=Table[l[[i,j]] *
(KwaveSquare[[i]]+KwaveSquare[[j]]),{i,n},{j,n}]

InverseLaplacien[l_,alpha_,beta_]:=Table[l[[i,j]] /
(beta(KwaveSquare[[i]]+KwaveSquare[[j]])+alpha),{i,n},{j,n}]

Nonlinear[l_]:=InverseLaplacien[PhysicalToSpectral[
 SpectralToPhysical[yDerivate[l]] *
 SpectralToPhysical[xDerivate[Laplacien[l]]] -
 SpectralToPhysical[xDerivate[l]] *
 SpectralToPhysical[yDerivate[Laplacien[l]]]],
 10^-40,1.]

EulerStep[l_]:=InverseLaplacien[l+dt Nonlinear[l],1,-nu dt]
```

Listing 1: Fourier transforms and numerical kernel coding.

and the FFTs are specially adapted to the machine's architecture (often written in assembly language).

Results from DNS of 2D turbulence have included the transition from an early-time k^{-4} inertial range to a latter time k^{-3} range, for flows developing from large-scale initial data[3]. An important finding has been the identification of the spontaneous emergence of long-lived coherent vortices when the DNS is run for a long time[4]. Long-time behavior of 2D turbulence is an active field of research, using DNS with a resolution of up to 1024^2[5].

The biggest general periodic computations to date in 3D turbulence have a resolution of up to 512^3 (performed on a CM5 massively parallel machine)[6, 7]; 256^3 computations are common on Cray supercomputers[8]. The Taylor-Green vortex, a periodic flow with symmetries that can be used to save time and memory, has been simulated with resolutions up to 864^3 on a Cray2[9, 10].

When studying 3D turbulence, the basic limitation of DNS is their lack of resolution. From Kolmogorov scaling one knows that the dissipation scale is $\ell_d \sim \ell_I R_I^{-3/4}$, where ℓ_I is the injection length scale and R_I is the Reynolds number. To accommodate this range of scales the DNS's resolution must scale as $n \sim R_I^{3/4}$, which implies computer memory scaling as $R_I^{9/4}$ and a total operation count scaling as $R_I^3 \log(R_I)$[11]. State-of-the-art 3D DNS, in

```
(*Visualizations*)
ShowMeOmega[l_]:=

ListContourPlot[Re[SpectralToPhysical[-Laplacien[l]]]]

(*Initialization*)
SeedRandom[143];
Field=Table[1/(
 -KwaveSquare[[i]]-KwaveSquare[[j]]+0.001) *
 ( ( Random[]-.5)+I (Random[]-.5)/2 ),{i, n}, {j, n}];
Field[[1,1]]=0.;
Psi=PhysicalToSpectral[Re[SpectralToPhysical[Field]]];

Do[Psi=EulerStep[Psi];Psi[[1,1]]=0.,{200}];
ShowMeOmega[Psi]
```

Listing 2: Initialization, time loop and visualization coding.

```
ShowMeDistribution[l_]:=ListPlot[
 Transpose[{Sort[Re[Flatten[l]]],
 Table[N[(i-1)/n^2],{i,n^2}]}]]
ShowMeDistribution[
 SpectralToPhysical[xDerivate[Laplacien[Psi]]]]
```

Listing 3: Code to compute the distribution function of $\partial_x\omega$.

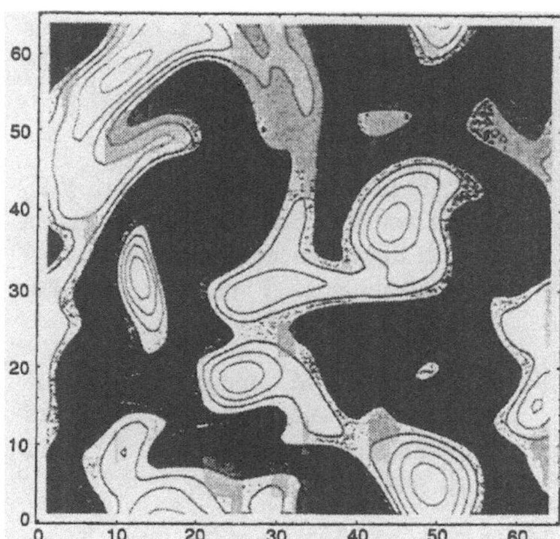

Figure 1: A plot of the vorticity generated by the toy solver.

the early 90's, typically have an R_l of a few thousands (and a Taylor-scale Reynolds number R_λ of a few hundreds). Given these limitations, why have DNS been so popular in the recent past? A first remark is that, with R_λ a few hundreds, one starts to see Kolmogorov scaling (over, say, a decade). So one can argue that the DNS has something to do with turbulence. It can be used to validate turbulence models[12]. Another point is the possibility of getting information about any computable quantity. Indeed, suppose we want to see a cumulative distribution plot (the integral of the p.d.f.) of e.g. the x-derivative of the vorticity field computed in our toy Mathematica 2D DNS. The 2 instructions displayed in List. 3 will produce the plot shown in Fig. 2.

This is indeed the very power of DNS: any quantities of interest can be computed, plotted and studied.

Evidence for the presence of vortex blobs, sheets, and filaments, in 3D DNS of turbulence, were first obtained in the early 80s[13]. At about the same time, the

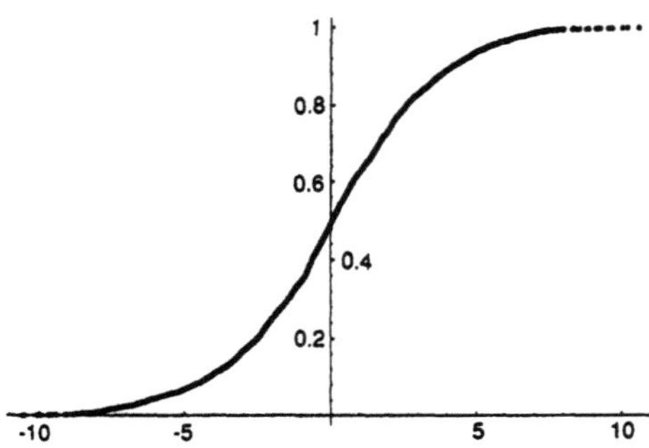

Figure 2: Cumulative distribution of the vorticity gradient (x-component) corresponding to the situation displayed in Fig. 1.

first Kolmogorov $k^{-5/3}$ inertial range was reported in a DNS[14]. It is impossible to give here an exhaustive list of all the important contributions that followed in the numerous papers published from the mid 80s to the early 90s. Perhaps two main results should be mentioned at this point, as they have both motivated new experiments that go beyond statistical analysis of one-point velocimetry data.

The first result is the DNS data on strain statistics and vorticity alignment with the eigenvector corresponding to its middle eigenvalue[15]. This result has been confirmed in sophisticated hot-wire measurements[16], that shed new light on the geometrical distribution of turbulent small scales. The second result is the utilization of DNS to gain insight about the coherent structures present in 3D turbulent flows and their physical signatures[9, 17]. These DNS have confirmed the presence of vortex filaments and sheets, and demonstrated the dominant effect of vortex filaments on the pressure field. Pressure field visualizations and histograms, obtained in the DNS have paved the way for new experiments that have directly visualized vortex filaments in real turbulent flows[18].

These new experiments have, in turn, focused the attention of the turbulence community on the study of small-scale geometry, filamentation and vortex breakdown,

as a basic dynamical mechanism in 3D turbulence.

Bibliography

[1] D. Gottlieb, S.A. Orszag, in *Numerical Analysis of Spectral Methods*, SIAM, Philadelphia (1977).

[2] The Mathematica source code, together with some extra functions needed to de-aliase, compute spectra, etc..., can be obtained by sending e-mail to: *brachet@physique.ens.fr*

[3] M.E. Brachet, M. Meneguzzi, P-L Sulem and H. Politano, The Dynamics of Freely Decaying Two-Dimensional Turbulence, J. Fluid Mech. **194**, 333 (1988)

[4] J.C. McWilliams, The emergence of isolated coherent vortices in turbulent flow, J. Fluid Mech. **146**, 21 (1984).

[5] V. Borue, Inverse energy cascade in stationary two-dimensional homogeneous turbulence, Phys. Rev. Lett. **72**,10, 1475 (1994).

[6] S. Chen, G. Doolen, R.H. Kraichnan,and Z.-S. She, On the statistical correlations between the velocity increments and the local dissipation rate, Phys. Fluid A **5**, 458-463 (1993).

[7] Z.-S. She, S. Chen, G. Doolen, R.H. Kraichnan and S.A. Orszag, Reynolds number dependence of isotropic Navier-Stokes turbulence, Phys. Rev. Lett. **70**, 3251-3254 (1993)

[8] A. Vincent and M. Meneguzzi, The spatial structure and statistical properties of homogeneous turbulence, J. Fluid Mech. **225**, 1 (1991).

[9] M.E. Brachet, Géométrie des structures à petite échelle dans le vortex de Taylor-Green, C.R.A.S II **311**, 775 (1990).

[10] M.E. Brachet, M. Meneguzzi, A. Vincent, H. Politano, and P-L Sulem, Numerical evidence of smooth self-similar dynamics and possibility of subsequent collapse for three-dimensional ideal flows, Phys. Fluids A 4, 12, 2845 (1992).

[11] U. Frisch, in, *Turbulence: The legacy of A.N. Kolmogorov*, to be published by Cambridge University Press (1995).

[12] B. Mohammadi and O. Pironneau, in *Analysis of the K-Epsilon turbulence model*, Research in applied Math., Wiley-Masson (1993).

[13] E.D. Siggia, Numerical study of small-scale intermittency in three-dimensional turbulence, J. Fluid Mech. **107**, 375 (1981).

[14] M.E. Brachet, D.I. Meiron, S.A. Orszag, B.G. Nickel, R.H. Morf, and U. Frisch. Small Scale Structure of the Taylor-Green Vortex, J. Fluid Mech. **130**, 411 (1983).

[15] R.M. Kerr, Higher order derivative correlations and the alignment of small-scale structures in isotropic numerical turbulence, J. Fluid Mech. **153**, 31 (1985).

[16] A. Tsinober, E. Kit and T. Dracos Experimental Investigation of the Field of Velocity Gradients in Turbulent Flows, J. Fluid Mech. **242**, 169-192 (1992).

[17] M.E. Brachet, Direct simulation of three-dimensional turbulence in the Taylor-Green Vortex, Fluid Dynamics Research **8**, 1-8 (1991).

[18] S. Douady, Y. Couder and M.E. Brachet, Direct observation of the intermittency of intense vorticity filaments in turbulence, Phys. Rev. Lett. **67**, 8, 983-986 (1991).

NUMERICAL SIMULATIONS OF TWO-DIMENSIONAL FLOWS
(Turbulence and vortices)

B. LEGRAS

Laboratoire de Météorologie Dynamique,
Ecole Normale Supérieure,
24 rue Lhomond, F-75231 Paris cedex 05, France

INTRODUCTION

As explained by P.Tabeling in this volume, fluid motion tends to be two-dimensional under the action of various factors (a) rotation, (b) stratification, (c) flat geometry and (d) action of a magnetic field. The first three conditions are met within the fluid envelop of the Earth where these properties force the fluid to move predominantly within strata, particularly at large scales. This motion is not purely two-dimensional since these strata are sloping surfaces and the baroclinic convection on these surfaces is actually the mechanism that converts potential energy into kinetic energy in the extra-tropical regions. Two-dimensional turbulence is, however, a useful model of the dynamics of established dynamical structures within the atmosphere and the oceans over a couple of weeks in the atmosphere and a few months in the ocean[1]. In the particular instance of the middle atmosphere between 20 and 40 km in altitude, where any convective motion is strongly inhibited by the thermal stratification, it is conjectured that two-dimensional motion can be observed down to horizontal scale of 50 m. This natural laboratory is unfortunately not easily accessible for high-resolution measurements. Satellite observations show us that the atmosphere of Jovian planets is also rich in behavior pertaining to two-dimensional turbulence.

The second historical motivation of studying two-dimensional turbulence has been the belief that it could serve as an intermediate model to understand the "real" three-dimensional turbulence or can serve as a test to the general statistical theories which were developed in an attempt to solve the turbulence problem. The underlying assumption was that two-dimensional turbulence should be much easier to understand and to modelize than three-dimensional turbulence. We know presently that this was a somewhat optimistic view.

It is however clear that numerical simulations are much easier in two dimensions than in three dimensions and can be performed at much higher Reynolds number. Numerical experiments of two-dimensional turbulence have been numerous since the early time of computer science. Owing to the difficulty of laboratory measurements and of in-field measurements, most of our understanding until recently came from these experiments.

GEOMETRY AND NUMERICAL TECHNIQUES

A favorite geometry of two-dimensional numerical experiments is a torus which is mapped onto a doubly-periodic plane in x and y.

Numerical calculations are then made easy by representing any field as a truncated Fourier expansion such as

$$\psi(x, y) = \sum_{k=-N}^{N} \sum_{l=-N}^{N} \hat{\psi}(k, l) e^{i(kx+ly)} \quad (1)$$

for the stream function. Spatial derivatives are calculated as products by constant coefficients in the Fourier space.

Solving the Poisson problem $\nabla^2\psi = \omega$ to calculate the stream function from the vorticity is easily done as $\hat{\psi}(k,l) = -\hat{\omega}(k,l)/(k^2 + l^2)$ owing to the double periodicity. The calculation of the non linear term, which leads to a convolution in Fourier space requiring $O(N^4)$ operations, is performed on a collocation grid in the physical space using Fast Fourier Transform to go back and forth between the two spaces, the number of operations being then reduced to $O(N^2 \log N)$. In order to preserve the basic truncation (1), one has to neglect the excitation of higher-order harmonics by the nonlinear term. Yet, the projection of the temporal derivative onto the represented modes can be calculated exactly by choosing a large enough collocation grid, thus minimizing the r.m.s. error of the evolution[2, 3].

A radically different method which applies in an unbounded domain is contour dynamics[4]. It consists in discretizing the vorticity fields by a finite set of contours and solving the Euler equation by evolving these contours under the action of the velocity field calculated at each point as an integral over the contours:

$$u(x) = -\sum_k \frac{\Delta\omega_k}{2\pi} \oint_{C_k} \ln|x - x_k|\, dx_k \quad (2)$$

where $\Delta\omega_k$ is the vorticity jump across the contour C_k. Contours are themselves discretized according to their length and curvature. The main interest of this Lagrangian method is to provide a self-adaptative resolution which crowds contours only where small-scale structures are generated by the dynamics and thus to reach spatial resolution unattainable by other methods. For far enough contours, an accurate approximation to the integral (2) is a much simpler sum over moments of the vorticity distribution which gives way to a very efficient algorithm able to handle hundreds of contours[5]. Contour dynamics is also a flexible tool which can be adapted to various geometries and to the existence of free-slip boundaries. For instance, it is not more costly to use it on a spherical surface than on a plane unlike the pseudo-spectral method. See also Ref. [6] for a detailed comparison of its accuracy with the pseudo-spectral method.

An other hybrid numerical method is the vortex-in-cell algorithm which represents vorticity as a sum of discrete vortices and solves the Poisson equation by averaging over a grid. See Refs. [7, 8] and references herein for a review of this approach.

A recent method of interest for flows with complex boundaries or in porous media is the lattice Boltzmann equation which calculates the evolution of packets of fictitious particles traveling on a grid at the nodes of which they collide and interact. See Ref. [9] for a thorough discussion of this method and Ref. [10] for a comparison with pseudo-spectral calculations.

Small-scale dissipation introduced in numerical simulations often differs from a standard molecular diffusion. Pseudo-spectral simulations use a hyperviscosity operator defined as an iterated Laplacian, that is written $D\omega/Dt = (-1)^n\Delta^n\omega$. Such mechanism generates a narrow dissipation range in wavenumber space, thus allowing to extend the range of inertial motion for a given resolution. In contour dynamics, the filamentary structures are selectively removed at very small scale. Notice that there is *a priori* no physical reason to use a standard horizontal diffusion since in practical situations, two-dimensional motion is halted by another mechanism such as friction, vertical diffusion or three-dimensional turbulence which, in addition, often acts at scales which are unresolved by the large-eddy numerical simulations. It is however necessary to understand whether the properties of two-dimensional turbulence are sensitive to the chosen parameterization. The general belief is that they are not for a given Reynolds number (defined here from the extension of the inertial range) but this question is not settled down. Recent calculations show that hyperviscosity can pump vorticity up intense vorticity gradients and generate spurious peaks in the vorticity distribution[11]. We also know that large-scale instabilities can be driven by dissipation and nonlinearities under a rather large set of circumstances[12]. Although this latter result is only established at low Reynolds number, it might be effective at higher Reynolds and introduce a dependency of the large-scale flow upon the dissipative mechanisms.

ENERGY SPECTRUM

Unlike three-dimensional motion, the conservation of vorticity for each particle implies that two-dimensional Euler equation possesses an infinite number of invariants, in addition to energy and total angular momentum. This conservation is lost under dissipative or forced conditions but is relevant for the motion at scales which are either large or small with respect to the excitation scale and large with respect to the dissipative scale.

Energy and enstrophy are both conserved within a basic nonlinear interaction between a triad of wavenumbers in phase space[13]. As a result emerges the notion of double cascade[14]: if the flow is excited at a given scale, enstrophy goes to smaller scales where it is dissipated while energy fills the large scales. This mechanism leads to the prediction of two inertial domains, the enstrophy and the energy inertial ranges within which the spectral energy density $E(k)$ scales respectively as k^{-3} and $k^{-5/3}$. An alternative view was presented by Saffman[15] who considered that sharp-edged distribution of vorticity is conserved by the dynamics yielding $E(k) \sim k^{-4}$ at large k; we shall see below that sharp vorticity gradients are indeed quite frequently generated by two-dimensional flows. In principle, power-law energy spectrum can result either from the tendency to develop singularities (k^{-3} and cusps, k^{-4} and fronts) in the inviscid limit or from the scaling properties of vorticity distribution[16, 17]. Recently Polyakov[18] proposed a new approach in which conformal field theory is used to solve the Hopf equations for the probability distribution of vorticity. This theory predicts anomalous scaling of the energy spectrum but, in its present state, allows many different solutions depending on external conditions.

Most of the numerical or laboratory experiments have been done for decaying turbulence for which the phenomenology of the direct enstrophy cascade is, to a large extent, common with the forced situation[19]. One main finding of those investigations[20, 21] is that the energy spectrum is significantly steeper than the k^{-3} prediction while a passive scalar transported by the flow still obeys this law[22]. The reason, to be detailed below, is the presence of numerous vortices which dominate the vorticity field and govern the nonlinear transfers.

Investigations of the energy inertial range which requires small-scale forcing have only been attempted in a few numerical investigations[23, 24]. The $k^{-5/3}$ is established after a rather short time and the subsequent evolution depends on boundary effects. Recent studies[24, 25] report, however, the formation of small-scale vortices which may modify the spectrum at long time.

COHERENT VORTICES

The major feature of two-dimensional flows, which is missed by the spectral analysis, is the spontaneous generation of numerous coherent intense vortices which dominate the large-scale motion[26, 27, 28]. By coherent, we mean here observable pattern occurring repetitively with a lifetime greatly exceeding its own internal circulation time-scale. Fig. 1 offers an example of self-organization into a system of interacting vortices in the case of a forced experiment. The vortices are also produced after an initial transient stage in decaying turbulence and, once there, dominate the rest of the evolution.

The reason of the spontaneous generation of vortices under a very large variety of circumstances is still unclear. An axisymmetric monotonous distribution of

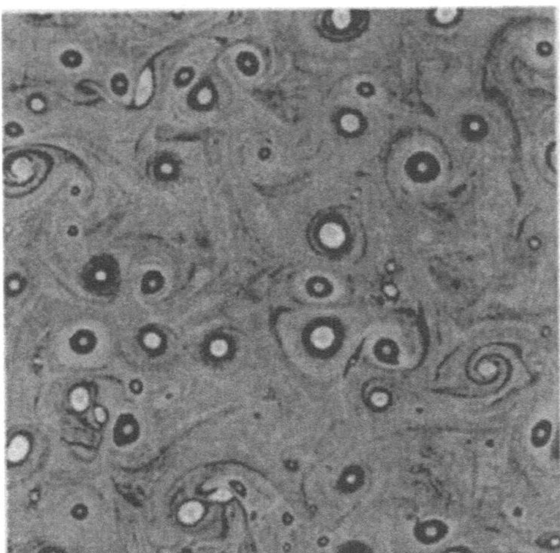

Figure 1: Vorticity chart from a numerical experiment of two-dimensional turbulence within a periodic box and excited by a large-scale instability of wavenumbers in the range $7 < k < 10$. We observe numerous vortices of various sizes and filaments between them being almost passively advected. Look at the various associations between vortices. One sees a tripole near the center and above it a vortex being torn apart by another one. Look also at the erosion and the generation of filaments around many of the vortices.

vorticity is known to be stable against finite amplitude perturbation. Non axisymmetric, quasi-elliptical structures can rotate and vacillate in a stable way as well. One possibility is that this stable distribution is selected as the only one which survives after a random exploration of the possible rearrangements of the initial conditions. The detailed examination of the early stage of a decay experiment from random initial conditions tends to support this hypothesis[20, 28]. If all coherent vortices can be traced back to initial maxima of the vorticity field, many maxima do not lead to a coherent vortex: they end into strained vorticity sheets and are dissipated[28]. A second possibility which *is developed in the article by Pomeau in* this volume is that turbulent mixing tends to generate a large-scale structure which maximizes its entropy, the small-scale de-

tails being in the meantime smeared out by dissipation. This seems to be relevant to the case of shear layers and to the final stage of a decay experiment. In general, the existence of a strong mixing is a prerequisite for this argument to apply. The third possibility is that vortices develop as the result of a large-scale instability over a turbulent flow. A famous example is the Kolmogorov flow[29] but it has been shown recently that this effect obtains for a large class of turbulent flows with isotropic small-scale forces[12].

Vortices are usually solitary but they quite often associate into bound states with two or more components. Some of these structures may survive for a very long duration in an unperturbed environment. They have been studied intensively both experimentally and theoretically (see Ref. [30] and references herein).

The space between the vortices in Fig. 1 is filled by filamentary structures which are submitted to the strain of the large-scale flow. As a consequence they undergo a series of stretchings and foldings, typical of the Hamiltonian chaos[31], decreasing the transversal scale until it reaches the dissipation scale. The large-scale strain is very effective in inhibiting the Kelvin-Helmholtz instability of the vorticity sheets which is only sparsely observed[32]. The contribution to the velocity field is small and these structures can be essentially considered as transported by the velocity induced by the dominating vortices.

It is thus tempting to see two-dimensional turbulence as the result of the collective dynamics of an ensemble of vortices. At first sight, vortices preserve their identity and a maximum simplification is to replace each of them by a point vortex located at its center with a charge equal to its total circulation. The charge does not change in time and each point vortex is advected by the velocity due to the other ones. The resulting system has a Hamiltonian structure and is known to generate chaotic motion when the number of vortices is four at least on the infinite plane or on the sphere and three on the periodic plane. The comparison of trajectories predicted by this approximation with those actually observed in a given numerical experiment exhibits a good agreement over a duration of about ten eddy turnover times[33]. The approximation ceases to be valid when a vortex undergoes an inelastic interaction with its neighbors.

INELASTIC INTERACTIONS

Merging

Unlike a couple of equal sign point vortices which rotates around their barycenter, finite vortices can merge when they come close enough. The simplest case is that of two initially identical circular patches of uniform vorticity. They merge into a single final vortex if the ratio between their initial distance and their diameter is less than 3.35[34]. The conservation of angular momentum implies that two filaments are ejected during the interaction which carry away the excess angular momentum while the final vortex essentially retains the total energy of the initial system. Dissipation is necessary in this process to smear out the fine scale spiral structure in the resulting vortex thus restoring an axisymmetric profile, but merging can nevertheless be considered as an inviscid mechanism. Merging of unequal vortices leads to a somewhat more complicated situation which allows partial merging or partial tearing of the weaker vortex[35]. As a result, the outcome is most often two vortices, with one being smaller and the other one being bigger than the two initial patches. Merging of two distributed vortices essentially follows the same rules. This phenomenon plays a major role in turbulent decay where bigger and bigger vortices are generated by successive merging leading to a final situation with only one pair of opposite sign vortices. The analysis of the evolution reveals that the process is governed by power laws in time. For instance the mean radius

follows $a(t) = t^\xi$ but the value of ξ is controversial since it differs from one numerical experiment to the other and in laboratory experiments as well. See the article by Y. Pomeau, P. Tabeling and W.R. Young in this volume.

Filamentation

A second inelastic mechanism — which has been largely overlooked in the literature on two-dimensional turbulence — is wave breaking on the periphery of vortices. Two phenomena are involved here. The first is the formation of nonlinear critical layers, to which classical analytical treatment may be applied to study the initial stages of perturbation development on a weakly unstable vorticity profile[36]. When a sharp vorticity interface is involved, the instability can be investigated using contour dynamics, and it can be shown[37] that a cascade of folding occurs which generates an extremely complicated structure of embedded filaments on the vorticity interface. There exist, however, nonlinear stability bounds which restrict the growth of such perturbations[38]. In particular, the mean-square displacement of the interface is conserved, and it is believed, on the basis of many numerical experiments, that the filaments remain close to the initial location of the vorticity interface. This phenomenon, known as "filamentation", is certainly important in determining an effective turbulent diffusion acting on the vorticity profile of a coherent vortex, but as just noted probably injects little material into the background flow.

Stripping

The second phenomenon, that we denote as "stripping", is associated with the deformation of a vortex by an external straining field produced by other vortices (or any imposed non-uniform flow). The superposition of a vortex having closed streamlines and an external straining field (i.e. $u = \gamma x - \Omega y, v = -\gamma y + \Omega x$)

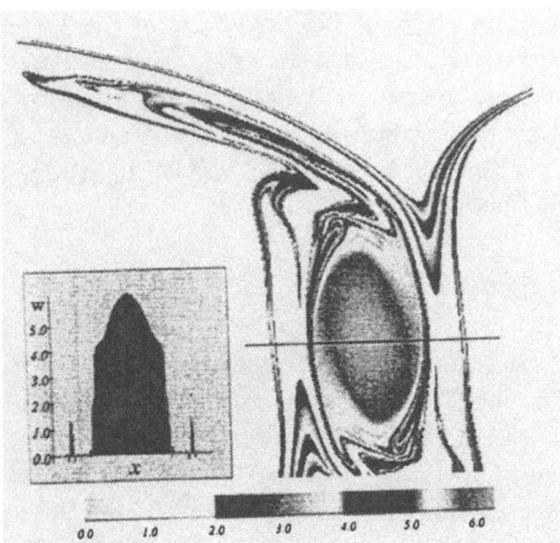

Figure 2: Vorticity field of a vortex submitted to a uniform adverse shear at a late stage after the stripping has started; filaments surround the vortex and a sharp boundary develops as is shown in the vorticity cross section.

yields a total stream function field having two critical points, often located inside the vortex, and associated separatrices passing through these critical points, capable of transporting fluid to large distances from the vortex into the background flow. For some combinations of γ and Ω, the vortex may be destabilized, resulting in tearing, folding and appreciable filament generation[39]. If the ratio of γ to the maximum value of vorticity within the vortex exceeds a certain critical value depending on Ω, the vortex breaks out. A noticeable consequence of partial stripping removing the periphery of the vortex is that the vorticity gradient increases dramatically on the edge of the vortex[40]. A sharp edge is indeed observed at the periphery of the polar vortex by in-situ aircraft measurements[41] and is found to be most pronounced following a strong perturbation of the vortex, in agreement with our analysis. In forced turbulence, stripping is the source which permanently injects new filamentary structures within the background flow where they are prone to Hamiltonian mixing until they reach the dissipative scale. Fig. 2 illustrates the ef-

fects of stripping near a vortex placed in a row of identical vortices.

In the absence of diffusion, stripping halts when the vortex reaches an equilibrium profile with the critical points located on its edge. When diffusion is coupled with stripping, we expect that there will be an accelerated decay of the vortex when one balances the diffusive flux of vorticity (accentuated by the presumed tight gradients along the vortex edge) across the separatrices of the stream function with the advection away by the external flow[42, 11]. We indeed observe that the vortex shrinks and that its vorticity maximum decreases linearly in time until it reaches the critical value of inviscid vortex breaking[11].

A practical consequence is that in large-eddy simulations which parameterize the unresolved scales by increased diffusion or hyperviscosity, one may expect from the above observations that the combined effect of eddy viscosity and strain lead both to a premature disappearance of small-scale eddies and to an enhanced erosion of large-scale structures[42], as conjectured previously in [43]. It is a mechanism which may explain why some coherent patterns like atmospheric blocking anticyclones are not properly maintained during the integration of numerical weather forecast models. The use of hyperviscosity may somewhat delay the final breaking since the maximum vorticity decays more slowly than in the case of normal diffusion. But there are also some spurious effects. Hyperviscosity is not a diffusive operator: artificial oscillations develop across a sharp edge leading to an up-gradient pumping of vorticity inside the vortex. This violates vorticity conservation, of course, and can enhance maxima by 20% above their initial value (or more if $p > 2$)[11]. The consequences of this result for forced simulations require further analysis. We can say, however, that it casts doubt on the statistics of small-scale vortices in experiments using hyperviscosity (see Ref. [43] for further remarks) and on the reported occurrences of cusp-shaped vortices in low-resolution experiments.

We remark that the ability of stripping to produce in one step and in approximately an eddy-turnaround time structures at the dissipation scale from initially large-scale vorticity gradients goes against the classical cascade concept. The latter assumes the existence of an inertial range in which the fluctuations are statistically independent of both the large-scale flow and dissipation. We see here that, on the contrary, we cannot assume any homogeneity or isotropy of the small scales since they are fully dependent on the large-scale flow. It remains to be seen whether the kind of Hamiltonian chaos we have encountered above is able to restore, at least partially, statistical independence for small-scale fluctuations. Any effort to restore the classical theory will, however, have to contend with a very recent result (R. Benzi, personal communication) that the Kolmogorov relation on third-order correlations, which can be derived exactly from the Navier-Stokes equations on the sole assumption of statistical homogeneity and isotropy, is *never* satisfied by known data on two-dimensional turbulence, unless conditions that forbid coherent vortices are imposed.

Bibliography

[1] J. Pedlosky, *Geophysical Fluid Dynamics*, 2nd ed. (Springer Verlag, New-York, 1987).

[2] G. Patterson and S. Orszag, Phys. Fluids **14**, 238 (1971).

[3] C. Canuto, M. Hussaini, A. Quarteroni, and T. Zang, *Spectral methods in fluid dynamics* (Springer, New York, 1987).

[4] D. Dritschel, Computer Phys. Rep. **10**, 77 (1989).

[5] D. Dritschel, Phys. Fluids A **5**, 173 (1993).

[6] B. Legras and D. Dritschel, J. Comput. Phys. **104**, 287 (1993).

[7] T. Sarpkaya, ASME J. Fluids Eng. **111**, 5 (1988).

[8] A. Leonard, J. Comput. Phys. **35**, 289 (1980).

[9] R. Benzi, S. Succi, and M. Vergassola, Physics Report **222**, 145 (1992).

[10] D. Martínez, W. Matthaeus, S. Chen, and D. Montgomery, Phys. Fluids **6**, 1285 (1994).

[11] A. Mariotti, B. Legras, and D. Dritschel, to appear in Phys. Fluids (1994).

[12] S. Gama, M. Vergassola, and U. Frisch, J. Fluid Mech. **260**, 95 (1994).

[13] R. Fjortoft, Tellus **5**, 225 (1953).

[14] R.H. Kraichnan, Phys. Fluids **10**, 1417 (1967).

[15] P. Saffman, Stud. Appl. Math. **50**, 377 (1971).

[16] R. Benzi, S. Patarnello, and P. Santangelo, J. Phys. A: Math. Gen. **21**, 1221 (1988).

[17] R. Benzi, G. Paladin, and A. Vulpiani, Physical Review A **42**, 3654 (1990).

[18] A. Polyakov, Nucl. Phys. B **396**, 367 (1993).

[19] G. Batchelor, Phys. Fluids Suppl. II **12**, 233 (1969).

[20] P. Santangelo, R. Benzi, and B. Legras, Phys. Fluids A **1**, 1027 (1989).

[21] B. Legras, P. Santangelo, and R. Benzi, Europhys. Lett. **5**, 37 (1988).

[22] A. Babiano, C. Basdevant, B. Legras, and R. Sadourny, J. Fluid Mech. **183**, 379 (1987).

[23] U. Frisch and P.-L. Sulem, Phys. Fluids **27**, 1921 (1984).

[24] L. Smith and V. Yakhot, Phys. Rev. Lett. **71**, 352 (1993).

[25] V. Borue, Phys. Rev. Lett. **72**, 1475 (1994).

[26] C. Basdevant, B. Legras, R. Sadourny, and M. Béland, J. Atmos. Sci **38**, 2305 (1981).

[27] J. McWilliams, J. Fluid Mech. **146**, 21 (1984).

[28] J. McWilliams, J. Fluid Mech. **219**, 361 (1990).

[29] G. Sivashinsky, Physica **17D**, 243 (1985).

[30] X. Carton and B. Legras, J. Fluid Mech. **267**, 53 (1994).

[31] J. Ottino, *The Kinematics of Mixing : Stretching, Chaos and Transport* (Cambridge U.P., Cambridge UK, 1989).

[32] D. Dritschel, J. Fluid Mech. **206**, 193 (1989).

[33] R. Benzi, S. Patarnello, and P. Santangelo, Europhys. Lett. **3**, 811 (1987).

[34] N. Zabusky, M. Hughes, and K. Roberts, J. Comput. Phys. **30**, 96 (1979).

[35] D. W. Waugh, Phys. Fluids A **4**, 1745 (1992).

[36] T. Warn and H. Warn, S.I.A.M. **59**, 37 (1978).

[37] D. G. Dritschel, J. Fluid Mech. **194**, 511 (1988).

[38] D. Dritschel, J. Fluid Mech. **191**, 575 (1988).

[39] D. Dritschel, in *Mathematical aspects of vortex dynamics*, edited by R. Caflisch (SIAM, 1989), pp. 107–119.

[40] B. Legras and D. Dritschel, Applied Scientific Research **51**, 445 (1993).

[41] D. Waugh *et al.*, J. Geophys. Res. (1994), to appear.

[42] H. Yao, D. Dritschel, and N. Zabusky, Phys. Fluids A (1994), to appear.

[43] D. Dritschel, Phys. Fluids A **5**, 984 (1993).

OPTICAL TURBULENCE

A. C. NEWELL
Arizona Center for the Mathematical Sciences,
Tucson, AZ 85721, USA
V. E. ZAKHAROV
Landau Institute for Theoretical Physics,
Chernogolovka, Russia

This essay is a "c'mon over" invitation to the battered veterans of the war to tame the turbulence realized by the whimsical and capricious solutions of the high Reynolds number limit of the Navier-Stokes equations. Simple fluids are easier to drink than understand. They admit no nontrivial limiting behaviors, no convenient footholds from which the theoretician can launch his first assault. When the flow is interesting, every term in the Navier-Stokes equation is equally important at some scale. Efforts to mimic the effects of scales below a certain size, beginning from the intuitive ideas of Prandtl about eddy viscosity, to the $K - \epsilon$ theories, to the renormalization approach, by defining bulk parameters to simulate small eddy influence on mean flow and transport properties are *ad-hoc* at best and misleading at worst. Indeed any theory that attempts to build a description of fluid turbulence solely on a division of the phase space into a basis directly correlated with length scale (e.g. Fourier transform) is probably doomed to failure because at best only certain eddy-eddy interactions can be summed to all orders and captured by perturbation or renormalization theories. Many of the most interesting structures realized by solutions of the fluid equations, e.g. tornadoes, hurricanes, filaments, are the result of a real but poorly understood confluence of substructures containing a whole spectrum of scales united in synchronous harmony. These ob-

jects are robust (the events are not rare and each event, once begun, is very stable) and almost singular solutions of the Navier-Stokes equations. They appear at random points at space and time, and are usually short lived. They, too, play a role in determining transport properties and dissipation rates. In particular, they contribute to the intermittency of turbulence, the bursts of large fluctuations that affect the tails of the joint probability density function for the velocity gradients and contribute to non-Kolmogorov type behavior of the moments of higher powers of velocity differences. To understand better the interplay of these two ingredients of turbulence, one based on the eddy picture and the other on the filament picture, and to gain some insight into possible new ideas for melding these two pictures into a coherent whole, we suggest that it is profitable to explore other field theories that allow one to separate these two ingredients in a more meaningful and quantitative way. In particular, we suggest that optical turbulence provides an ideal paradigm because its Kolmogorov type behavior can be captured by weak turbulence theory, the structure of its (almost) singular solutions is known and the reasons for their nucleation understood. Hence our invitation to *c'mon over* to a different paradigm for turbulence.

In this short essay, we will spell out what we mean by optical turbulence, and discuss some of the advantages, and then summarize in a table form a comparison between the turbulence of the Euler/Navier-Stokes equations and their optical counterparts. Most of the ideas contained here have been already published in a number of articles[1-3] and the interested reader should consult them for details.

OPTICAL TURBULENCE

Optical turbulence in general is the study of the properties of the propagation of light in nonlinear dielectrics with energy sources and sinks included. Here, however, we use the term in a much more restricted way. By optical turbulence, we will mean the behavior of a singly polarized and almost monochromatic beam of light in which the electric field perturbations in the direction transverse to the direction of propagation are small amplitude and long with respect to the wavelength of the underlying carrier wave. We are therefore looking at the redistribution of energy among the scales contained in the wave packet about $\vec{k} = (0, 0, k = n(\omega)\omega/c)$. The relevant mathematical model (modulo a few caveats) is the forced-damped, nonlinear Schrödinger (fd NLS) equation:

$$A_z + k'A_t - \frac{i}{2k}\nabla^2 A + \frac{ik''}{2}A_{tt}$$
$$-i\frac{n_2(\omega)\omega}{c}|A|^2 A = -\hat{\gamma} \cdot A \qquad (1)$$

where $k' = \partial k/\partial\omega$, the refractive index of the medium is $n + n_2|A|^2$, and the term on the RHS of (1) is a convolution representing the input and output of power. The Fourier transform $\gamma(k)$ of $\hat{\gamma}$ is supported and negative (input) on a band of wavenumbers $(k_0 - \nabla, k_0 + \nabla)$ in the middle of the spectral range and is positive (output) in a range of wavenumbers $(0, k_1)$ and (k_d, ∞) near the origin and at infinity. We call the intervals $(k_1, k_0 - \nabla)$ and $(k_0 + \nabla, k_d)$ the left and right windows of transparency because energy and particle number simply flow through them and are neither produced nor dissipated there. Since the dispersion in the propagation direction k''/t_0^2 (t_0 is pulse length) is usually

much smaller than diffraction $(kw_0^2)^{-1}$ (w_0 is beam width), we examine first the "long pulse" model and ignore t derivatives. The resulting two dimensional fd NLS has the following properties and advantages:

a. The conservative part ($\gamma = 0$) has <u>three</u> conserved quantities:
particle number

$$N = \int AA^* d\vec{x} = \int A_k A_k^* d\vec{k},$$

momentum

$$P = i/2 \int (A\nabla A^* - A^*\nabla A)d\vec{x} = \int \vec{k} A_k A_k^* d\vec{k},$$

and energy

$$H = \int (|\nabla A|^2 - \frac{1}{2}\alpha|A|^4)d\vec{x}$$
$$= \int \omega_k A_k^* A_k d\vec{k} +$$
$$\frac{1}{2}\int T_{kk_1,k_2k_3} A_k^* A_{k_1}^* A_{k_2} A_{k_3}$$
$$\delta(\vec{k} + \vec{k}_1 - \vec{k}_2 - \vec{k}_3)d\vec{k}d\vec{k}_1 d\vec{k}_2 d\vec{k}_3,$$

where $\omega_k = k^2$, $T_{kk_1,k_2k_3} = -\alpha/(2\pi)^2$, $\alpha = n_2\omega/c$ and A_k is the Fourier transform of A.

b. For small amplitudes and for the power (particle number, energy) located at moderate wavenumbers, the Hamiltonian is dominated by its quadratic part and the equation is dominated by its linear term. In particular, <u>weak turbulence</u> theory applies and a kinetic equation for the spectral energy $n_k, \langle A_k^* A_{k'} \rangle = \delta(\vec{k} - \vec{k}')n_k$ obtains:

$$\frac{\partial n_k}{\partial z} + 2\gamma_k n_k = st(n, n, n) =$$
$$4\pi \int |T_{kk_1,k_2k_3}|^2 \delta(\vec{k} + \vec{k}_1 - \vec{k}_2 - \vec{k}_3)$$
$$\delta(\omega + \omega_1 - \omega_2 - \omega_3)$$
$$(n_{k_1}n_{k_2}n_{k_3} + n_k n_{k_2}n_{k_3}$$
$$-n_k n_{k_1}n_{k_3} - n_k n_{k_1}n_{k_2})d\vec{k}_1 d\vec{k}_2 d\vec{k}_3. \qquad (2)$$

In deriving (2), no closure assumptions are required. To leading order, the third and higher order cumulants all contain fast oscillations $\exp i(\omega + \omega' + \cdots)t$ and the fully dispersive nature of the system leads to a resonant manifold

$$\vec{k} + \vec{k}' + \cdots = 0$$
$$\omega + \omega' + \cdots = 0 \qquad (3)$$

on which $\nabla_{\vec{k}}(\omega + \omega' + \cdots) \neq 0$. The assumption of spatial homogeneity means that cumulants such as $\langle A(x)A(x + r_1)A(x + r_2)A(x + r_2)\rangle - P_{123}\langle A(x)A(x + r_1)\rangle\langle A(x + r_2)A(x + r_3)\rangle$ depend only on the relative geometry and so statistical averages can be found by averaging over the base vector $\vec{x}$. Then $P \equiv 0$. We are in fact going to study for pedagogic reasons the equation (2) for general T_{kk_1,k_2k_3}. The reason for this is that the kinetic equation for fd NLS has only marginally (and in some cases, not at all) converging Kolmogorov solutions.

THE KINETIC EQUATION

The kinetic equation clearly identifies the mechanism for spectral energy transfer as a <u>four-wave resonance</u>. It provides a concrete realization of what we referred to as the eddy picture. Moreover, in addition to the thermodynamic equilibria $n_k = \frac{T}{\mu + \omega_k}$ (T the temperature, μ the chemical potential), which carry no flux of particle number or energy, it admits Kolmogorov spectra[4] corresponding to finite fluxes of <u>energy</u> to high wavenumbers through the right window of transparency $(k_0 + \Delta, k_d)$ where it is absorbed by the sink in (k_d, ∞) and <u>particle number</u> to low wavenumbers through the left window of transparency where it is either absorbed by the sink in $(0, k_1)$ or produces collapsing filaments. The existence of the inverse cascade of particle number via a combination Kolmogorov-thermodynamic solution was established in Ref. [3] by direct numerical simulations.

THE KOLMOGOROV SOLUTIONS

Assume $\omega = k^\alpha$ and that T_{kk_1,k_2k_3} is homogeneous of degree β, then the pure Kolmogorov solutions of (2) are:

$$n_k = c_1 Q^{1/3} k^{-\frac{2\beta+3d}{3} - \alpha/3}, \qquad (4)$$

$$n_k = c_2 P^{1/3} k^{-\frac{2\beta+3d}{3}}. \qquad (5)$$

For (4), Q is the flux of particle number n_k towards low wavenumbers, and $P = 0$.

For (5), P is the flux of energy $\omega_k n_k$ towards high wavenumbers and $Q = 0$. The Kolmogorov constants c_1, c_2 can be explicitly calculated. The existence of the solutions depends on the existence of certain integrals (the analogue of the locality hypothesis of Kolmogorov) and the positivity of P and Q. The adjective *pure* connotes solutions that have only one flux and are independent of temperature. There is no reason, however, to preclude solutions $n_k(k, T, \mu, P, Q)$ which depend on all parameters. Indeed, such solutions may be more relevant for fd NLS (see Ref. [3]).

The point we stress, however, is that in weak turbulence theory, the existence and stability of Kolmogorov solutions are concrete realities in which the notions of local transfer, dependence of n_k on P, Q etc. make quantitative sense. They are consistent with dimensional analysis. Multiply $st(n, n, n)$ by k^{d-1} and $\omega_k k^{d-1}$ respectively, equate to $\partial Q/\partial k$ and $\partial P/\partial k$ respectively and use dimensional analysis to get the Q, P and k dependence in (4), (5). They are consistent with the Kolmogorov hypotheses. But they are realized as attracting solutions of the kinetic equation. There is an additional important point. Observe that the convergence of the energy integral $\int \omega_k n_k d\vec{k}$ at $k = \infty$ on the Kolmogorov solution (5) depends on whether

$$\alpha - \frac{2\beta}{3} < 0 \qquad (6)$$

or not. If (6) holds, we say the system has <u>finite capacity</u>. This is the case for the Kolmogorov $k^{-5/3}$ spectrum in fully developed hydrodynamic turbulence. The question in this case is: how can one continue to pump energy into the system governed solely by the Euler equations and maintain energy conservation? On the other hand, if the opposite holds, we say the system has <u>infinite capacity</u>. In that case, it can, in principle, continue to absorb the injected energy indefinitely. The "optical" turbulence situation (for general T_{kk_1,k_2k_3}) provides some understanding on this dilemma. One can find[3, 4] the following self similar solution to the time dependent but dissipation free kinetic equation (2),

$$n_k = \frac{1}{z^a} n_0 \left(\frac{k}{z^b} \right) \qquad (7)$$

where $a = \frac{3d+2\beta}{3\alpha-2\beta}, b = \frac{3}{3\alpha-2\beta}$. The values for a and b are found directly from (2) and by assuming that the energy $\int \omega_k n_k d\vec{k}$ increases linearly in time. Observe that (7) corresponds to a front in k space traveling as

$$k \sim z^b = z^{\frac{3}{3\alpha-2\beta}}. \qquad (8)$$

Behind the front, the solution

$$n_0 \sim c \left(\frac{k}{z^b} \right)^{-(2\beta+3d)/3} \qquad (9)$$

exactly matches the Kolmogorov solution (5) as the z dependence in (7) cancels. This solution describes how the Kolmogorov solution is realized as the time independent wake of a front which travels to $k = \infty$. Now, in the infinite capacity case, $b = 3\alpha - 2\beta > 0$, the front gets to $k = \infty$ after an infinite time z so that the energy is balanced by the inviscid equations for all time z. In this case, also, because of the divergence of the energy integral, even for the Kolmogorov solution, most energy is contained in the small scales.

For the finite capacity case, all is different. First of all, most of the energy is contained in the large scales so that energy and dissipation spectra have widely separated peaks in k space. Second, the front (7) gets to $k = \infty$ in a finite time z_0 (here $z_0 = 0$) because $b = 3\alpha - 2\beta < 0$ and energy is only strictly conserved up to that time. What happens when $b < 0$ is that the four wave resonance carries energy very quickly to infinity and sets up very complicated (possibly) fractal solutions of the governing inviscid equations, behavior which is immediately regularized when one adds viscosity at these high wavenumbers. This is the case analogous to three dimensional hydrodynamic turbulence. In hydrodynamic turbulence, of course, the solution corresponding to (7) is no longer explicit but we surmise that the eddy-eddy interactions for a fluid system obeying the

Kolmogorov hypotheses set up the $k^{-5/3}$ solution in a similar fashion.

THE INVERSE CASCADE

The presence of an inverse cascade of particle number is a direct consequence of the conservation of both particle number and energy. This inverse cascade plays a very important role for fd NLS in that it provides the seed for the destruction of weak turbulence theory and forces us to deal with the nucleation of new structures which depend on a direct balance of linear and nonlinear terms.

The argument for an inverse cascade is as follows. Suppose n_0 particles are added in $(k_0 - \Delta, k_0 + \Delta)$. Their total energy is $n_0\omega_0$. If the net flow of energy is to higher wavenumbers, conservation of energy through the window $(k_0 + \Delta, k_d)$ means $n_0\omega_0 \simeq n_d\omega_d$ and, since $\omega_d \gg \omega_0, n_d \ll n_0$. Conservation of particle number, then, means that the excess particles must flow to lower wavenumbers. But, as particle number accumulates at lower wavenumbers, the Hamiltonian is less and less dominated by its quadratic term and indeed the two terms become comparable. The weakly nonlinear problem becomes fully nonlinear.

WHICH SOLUTIONS DOES THE FULLY NONLINEAR NLS ACCESS?

While the behavior of solutions of the kinetic equation is independent of the sign of α, the fully nonlinear behavior is not. For $\alpha > 0$, the ubiquitous self-focusing, modulation or Benjamin-Feir instability, seen in optics, plasmas and water waves, obtains. The reason is simple. Imagine a constant intensity, uniform in x and y, electric field wave propagating in the z-direction. Perturb the plane (x, y) of constant electric field so that in some regions the intensity is increased and in some it is decreased. The light rays will bend towards the former regions and away from the latter. The intensity in the former regions increases, the refractive index becomes even larger there, inducing more focusing of the rays and so on until diffraction opposes the tendency. The "balance" between nonlinear focusing and diffraction in one dimension leads to the formation of stable solitons. In two dimensions, it leads to a particular and well defined soliton-like shape, the collapsing filament, whose amplitude increases like $\tau(z)(z_0 - z)^{-1/2}(\tau(z)$ is slowly varying), whose width decreases in inverse proportion and whose local wavenumber increases as $(z_0 - z)^{-1}$. Each of these filaments carries a constant amount of particles number N_c. These collapsing filaments have been known to light scientists for thirty years. They are known to damage material by local overheating and recently have been shown to be responsible for eye damage by moderate power but ultra-short pulses. These are the filaments of optical turbulence. Their structure and "interaction" behavior does not in any way satisfy Kolmogorov hypotheses. Their three dimensional cousins are even more dramatic and will be discussed later.

At this stage, we can describe, at least in qualitative terms, the cycle of intermittency.

THE CYCLE OF INTERMITTENCY

We carried out the following experiments. In the first, we set $\alpha < 0$, the defocusing case. There was clearly an inverse cascade but it was not of pure Kolmogorov (as in (4)) type. Rather it was a mixture of a thermodynamic and finite flux solution, and is described in Ref. [3]. The result was the building of a condensate at $k = 0$. Without damping, the amplitude of the condensate was observed to increase linearly with time representing a constant flux of particle numbers towards $k = 0$. With damping, the amplitude saturated. Phase differences at different locations can give rise to domain walls and defects but we will not discuss these further here. No intermittency was observed.

Next, we took $\alpha > 0$ and applied moderate damping in the low wavenumber sink $(0, k_1)$. We observed sporadic and infrequent intermittency. Next, we gradually removed the damping altogether, and observed a more and more intense outburst of intermittent behavior. Each large fluctuation occurred as a random event in time and space[†], and consisted of a collapsing filament. This filament had exactly the shape of the collapsing singular solution spoken of earlier. Each is nucleated by a modulational instability. The growth rate of the modulational instability depends on *enough* particle number being at sufficiently low k. Near $k = 0$, the notion of enough can be precisely stated. We need $\int |A|^2 d\vec{x} = N_c$. For a more complicated superposition of almost periodic wavetrains, the condition is more difficult. We return to this point in the next section. In this way, through randomly occurring collapse events, particles are carried back to large wavenumbers by a direct transport. The particle number is simply squashed from large scales to small scales as the filament gets larger and thinner. Because each filament carries the exact amount of particle number N_c needed to sustain collapse, the collapse is arrested when dissipation goes to work and reduces N below N_c. The amount of N lost per event, per burnout, is therefore less than N_c, usually about 15-20%N_c. Nevertheless a significant amount of the total dissipation of particle number is achieved through this secondary transport mechanism. The dissipation rate, then, in optical turbulence, depends on a cascade-like "local" transfer which satisfies the Kolmogorov hypotheses and a nonlocal direct transfer via randomly occurring coherent structures, the collapsing filaments.

What is important to note is that the universal constants and behaviors of each process may be quite different and somewhat independent. The Kolmogorov behavior is captured by certain statistical properties of the electric field itself, whereas the intermittent behavior depends more on variables, such as local dissipa-

---

[†]the introduction of delay or memory in the refractive index dependence on intensity can lead to a correlation between the spatial location of events

tion, which exhibit large fluctuations from one location to another and from one time to another.

A CHALLENGE: A QUANTITATIVE MODEL FOR THE CYCLE OF INTERMITTENCY.

We would like to construct a complete model which simulates in a quantitative manner this proposed cycle of intermittency. A natural decomposition of the field should contain two components, <u>the wave field</u>, the mutual interaction of whose spectral components is described by the kinetic equation, and the <u>filament field</u>. The statistics of the former field are likely to be close to Gaussian. The statistics of the latter field are likely to be Poisson in time and uniform in space with the average frequency per event a function of the rate at which the modulational instability can be triggered. This is an important point in that it has consequences for singular collapse production in the other fields of turbulence. Therefore, we will elaborate in a little more detail.

The inverse cascade of particle number is certainly, as we have demonstrated in Ref. [3], a <u>sufficient</u> mechanism for the production of filaments. The more the particle number is pushed back towards $k = 0$, the easier it is to cross the modulational instability threshold. However, the inverse cascade is not necessary for the formation of filaments. Indeed if the pumping occurred in an extremely narrow band, $\Delta \ll 1$, and at a sufficiently fast rate, then the wavetrain at k_0 could become modulationally unstable before or on the same time scale on which the four wave interaction process spreads the energy. In that case, filaments could nucleate and carry the particle number to the small dissipative scales in parallel with the Kolmogorov cascade. No inverse cascade would be necessary. The inverse cascade is simply added insurance that the collapse events will happen, sooner or later.

Therefore one would like to be able to calculate, given the pumping rate and the width of the pumping window, namely $\gamma(k)$ between $k_0 - \Delta$ and $k_0 + \Delta$, the frequency of collapse events. Each event will occur over a very short time and have the net effect of depositing a fraction f of N_c into the dissipation sink. The incomplete burnout will lead to a new source of the wavefield, as $(1-f)N_c$ of the particle number get reintroduced into the wavefield at wavenumbers typically between k_0 and k_d. These particles again will drift to lower wavenumbers and be part of new filament formation. They live to fight and die another day!

For this goal, as the reader will have no doubt noted, the fd NLS as it stands is not completely satisfactory because (i) the Kolmogorov finite energy flux solution $n_k \sim P^{1/3} k^{-d}$ has a divergent energy so that most of its energy is contained in short scales (some call this anti-Kolmogorov behavior but that is probably a bad name), (ii) there is no known exact mixed thermodynamic-Kolmogorov solution for the inverse cascade.

CONSEQUENCES FOR OTHER FIELDS:

There are many situations, ocean waves, plasma waves, atmospheric waves in which excursions from some equilibrium state are opposed by linear forces, for which weak turbulence theory obtains. However, it does not necessarily obtain throughout the whole range of wavenumbers, nor if a certain band of wavenumbers is pumped at too fast a rate. Therefore, either by inverse cascades which carry particles to wavenumbers where the Hamiltonian is no longer dominated by its quadratic part, or by direct forcing, certain instabilities can be triggered. In phase space, these saddle points can be connected with other equilibria in the finite part of phase space or infinity. If the unstable manifolds connect to infinity (and these connections in general are rare and difficult to make), they represent very well defined, coherent and robust collapsing structures just like the filaments in two dimensional nonlinear optics. These shapes would describe the bursting events.

Property	Euler/NS	Optics
Transport properties: Heat, momentum transfer Energy transport in k-space (dissipation rate)	$K - \epsilon$ theories *ad-hoc*	Open
Closure Origin of irreversibility	No ?	Yes, for finite but long times $\checkmark$
Solutions: Thermodynamic Kolmogorov Mixed type Decay of nonforced turbulence	? dim. argument ? late times only	$\checkmark\checkmark$ $\checkmark\checkmark$ $\checkmark$ $\checkmark$
Role of motion invariants in inverse cascades	$\checkmark$	$\checkmark$
Breakdown to filaments Mechanism of formation Demise of filaments	? ? some useful experimental and numerical work	$\checkmark$ $\checkmark$
Cycle of intermittency	some recent progress cf. She	qualitative

But hydrodynamics may also be more complicated than optical turbulence in that there could be a range of "collapsing" filaments. Velocity gradients may collapse as $(t_0 - t)^\gamma$ for a continuous set γ – in contrast to the NLS where $A \sim (z_0 - z)^{-1/2}$ (well, almost!). In such a scenario, one would have to have a means of determining an appropriate measure on this set to describe the frequency of an event with a particular γ occurring.

Even though fully developed, three dimensional Euler turbulence does not contain a linear restoring force, it is remarkable that again the turbulent field seems to exhibit a strong Kolmogorov character. In other words, in the absence of collapse events, or in situations where they are rare, the eddy picture seems to obtain. However, for higher and higher Reynolds numbers, it would appear that there are certain instabilities which lead to collapse events. The conditions for the formation of these saddles are not known. We have made some conjectures[1, 2]. What should

be clear is that they can be forced directly if the forcing is applied in the "sensitive" range of scales or by inverse cascades if the "sensitive" range is at larger scales. This scale L should play an important role in the theory. It may be roughly equal to the container size or the Kolmogorov size but it is not these sizes which matter. What is also clear is that the statistical behavior of quantities such as the local vorticity squared or the local dissipation rate should be relatively independent of the statistical properties of the velocity gradients themselves because the fluctuations are so localized. This realization is the basis of much of the recent work of She *et al.*[6] and is consistent with our picture.

We finish with two further remarks. The first concerns three dimensional, self-focusing fd NLS. In this case, there is a new kind of collapse event, a black hole which can be relatively long lived and absorb particle number from its neighboring field over finite time periods[5]. 2D filaments give up their particle number quickly. There is no analogue of these singularities in the three dimensional Euler equations. Perhaps in four dimensions! Second, we mention the interesting problem of sound turbulence where $\omega = c|\vec{k}|$. It is relatively easy to establish that the kinetic equation (with three wave resonances) obtains and redistributes energy along wavevector rays in $\vec{k}$ space. It is not at all clear, yet, what mechanisms move the energy from ray to ray. The turbulence of sound, however, may be an important intermediate case between the usual weak and fully developed turbulence behaviors.

A COMPARISON

See Figure at the top of the previous page.

Bibliography

[1] E. Kutnetsov, A.C. Newell and V.E. Zakharov, *Intermittency and Turbulence*, Phys. Rev. Letters **67** (23), 3243-3246 (1991).

[2] S. Dyachenko, A.C. Newell, A. Pushkarev, and V.E. Zakharov, *Optical Turbulence, Condensates and Collapsing Filaments in the Nonlinear Schrödinger Equation*, Physica D **57**, 96-160 (1992).

[3] A.C. Newell, *Inverse Cascades*, Physica D **61**, 213-216 (1992).

[4] V.E. Zakharov, V. L'vov and G. Falkovich, *Kolmogorov Spectra in Turbulence I*, Springer-Verlag (1992).

[5] N.E. Kosmatov, V.F. Shvetz and V.E. Zakharov, *Computer Simulation of Wave Collapses in the Nonlinear Schrödinger Equation*, Physica D **52**, 16-35 (1991).

[6] Z.-S. She and E. Leveque, *Universal Scaling Laws in Full Developed Turbulence*, Phys. Rev. Letters **72**, 336-339 (1994).

Phase Turbulence

H. Chaté and P. Manneville
CEA, Service de Physique de l'Etat Condensé,
Centre d'Etudes de Saclay,
F-91191 Gif-sur-Yvette, France
LadHyX — Laboratoire d'Hydrodynamique,
Ecole Polytechnique,
F-91128 Palaiseau, France

Phase Turbulence: this expression is used to designate the spatiotemporally disordered evolution of a phase variable in an extended system. It has little to do with the fully developed turbulent regimes produced by the Navier-Stokes equations. It was first used in the context of *phase equations*, i.e. partial differential equations to which the essential dynamics of some systems can be reduced, past a continuous-symmetry breaking instability threshold. We first describe this approach and its underlying assumptions, before detailing the particular but ubiquitous case of the complex Ginzburg-Landau equation and its associated phase equations.

GENERAL CONTEXT: PHASE EQUATIONS

The setting is that of an autonomous, spatially extended system evolving in a homogeneous medium which undergoes a cellular ($k_0 \neq 0$) and/or oscillatory ($\omega_0 \neq 0$) instability at some "critical" control parameter value. For simplicity, the system size is often taken to be infinite or at any rate large enough so that boundary effects can hopefully be treated as perturbations of the infinite-size case. Near the instability threshold, a local variable in the bifurcated state is then typically written[†]:

$$U(x,t) = U_0 + A \exp[i(k_0 x - \omega_0 t)] + c.c.$$

[†]In the following, a one-dimensional case is implicitly assumed for convenience but the procedure can be applied to higher dimensional cases. When $k_0 \neq 0$ and $\omega_0 \neq 0$, a case corresponding to the ap-

where U_0 is the "reference" state and A is a complex amplitude. The bifurcated solution ($A \neq 0$) breaks the continuous space and/or time translation invariance of the reference state. The translation invariance of the governing system imposes that any solution translated ($x \rightarrow x_0 + x$, $t \rightarrow t_0 + t$) from U remains valid. This, in turn, is transformed into the *gauge invariance* of the amplitude A; A is defined up to an arbitrary *phase* ϕ_0: $A \rightarrow A \exp i\phi_0$ with $\phi_0 = k_0 x_0 - \omega_0 t_0$.

From the point of view of the linear stability analysis of the bifurcated solution U, the translation invariance of the governing system is equivalent to the *neutral* character of the mode associated with homogeneous phase perturbations of A. Moreover, in an infinite, homogeneous medium such as the one considered here, this neutral mode belongs to a continuous branch parametrized by the wavenumber q of the phase perturbations (with the growth rate

pearance of waves at the instability threshold, one should, in principle, use two amplitudes A_L and A_R to account for the "left" moving and "right" moving waves: $U = U_0 + A_L \exp[i(k_0 x - \omega_0 t)] + A_R \exp[i(k_0 x + \omega_0 t)] + c.c.$. In practice, it is often the case that one direction is preferred, and that only one amplitude is needed, either because the system is not parity-invariant (e.g. open flows) or because the stable bifurcated solution selected involves only one set of waves (as in convection in binary mixtures[3, 4]). However, situations where $A_L = A_R \neq 0$ (stationary waves) or even $A_L, A_R \neq 0$, $A_L \neq A_R$ may arise, and one is then led to a system of two coupled first-order phase equations.

$\sigma(q)$ of these modes going to zero for homogeneous perturbations: $\lim_{q\to 0}\sigma(q) = 0$). Meanwhile, modes associated with other types of perturbations (e.g. perturbations of $|A|$) are typically strongly stable and quickly damped.

The basic idea behind what is sometimes called phase dynamics is that the evolution of the solutions close to U is confined to the *center manifold* spanned by the quasi-neutral, small-wavenumber phase perturbation modes. Along this manifold, the phase ϕ is slowly modulated in space and/or time ($\phi_0 \to \phi(x,t)$), and thus the phase gradients remain small. The procedure requires a legitimate separation of these phase modes from the other, "slaved" modes. It usually involves an expansion of the solutions close to U in powers of the phase gradient, which is then plugged into the governing system. At a given order of truncation of the expansion, the compatibility condition takes the form of an equation involving only the phase gradients:

$$\partial_t\phi = \mathcal{F}(\partial_x, \partial_x\phi).$$

In principle, the only assumption made is that the phase gradients remain small within the derived phase equation, which has to be checked *a posteriori*.

In many cases, the bifurcated solution U is not known explicitly, but only *via* an expansion in powers of the (small) distance ε to the instability threshold. Phase equations can then be obtained either from a double expansion in ε and the phase gradients, or from a two-step approach which consists in, first, deriving an *amplitude equation* governing the large-scale modulations of A (by an expansion in ε), and then performing a phase-gradient expansion on the phase-winding (constant $|A|$) solutions of this amplitude equation corresponding to the bifurcated solution U of the original system[†]. The terms of the

---

[†] Problems may arise along the two-step approach as the amplitude equation is usually valid only in the immediate vicinity of the bifurcation threshold.

phase equations reflect the invariances of the original problem. This is so important that in some cases, a phase equation can be derived only by symmetry considerations. Of course, the efficiency of this third method is balanced by the lack of information about the coefficients of the various terms, which have to be determined by other means. In particular, the coefficients of the linear terms can be given by the expression of $\sigma(q)$ (provided it is explicitly known). A classical example of great importance is the case of a one-dimensional system invariant by parity ($x \to -x$). This symmetry imposes that the lowest-order phase equation reads:

$$\partial_t\phi = D\partial_{x^2}\phi + g_0(\partial_x\phi)^2 \qquad (1)$$

where D is a *phase diffusion*[2] coefficient of unknown sign. If $D > 0$, phase perturbations do diffuse away, and the gradients remain small. This is the case when the solution is linearly stable against small-wavenumber perturbations ($\sigma(q) < 0$ for small q) since $D = -\frac{1}{2}\partial\sigma(q)/\partial q^2|_{q=0}$. Whenever $D < 0$, the long-wavelength phase modulations grow exponentially. To compensate for this *phase instability* (at least at the linear level), the phase equation has to be supplemented by higher-order terms, in particular possibly stabilizing linear terms, with nonlinear terms ensuring the transfer of spectral power from small to large wavenumbers.

In fact, nothing ensures that the phase equation at a given order will be "well-behaved", i.e. that it will exhibit a dynamics in which the phase gradients remain small, or even that instabilities or finite-time singularities will not occur. For the sake of completeness, the possibility of more complicated situations has to be mentioned. As already noted, in some cases several phases may be needed to specify the bifurcated solution U. One is then led either to a system of coupled phase equations (as in the case of left-going and right-going waves) and/or to equations with higher-order time derivatives (this happens in the specific but important case where several continuous symmetries are simultaneously broken at the bifurcation point).

At any rate, the procedure sketched above leads to partial differential equations governing the phase(s). *Phase turbulence* was first coined to describe the chaotic regimes that these phase equations exhibit in some cases. As for the relevance of this dynamics to the original problem, it should in principle be checked against the assumptions made when deriving the phase equations (i.e. essentially the smallness of the phase gradients). In practice, many efforts have been devoted to studying the dynamics of phase equations irrespective of the original systems they are derived from. Furthermore, the complex dynamics of some partial differential equations which, in another context, appear as phase equations, are sometimes wrongly depicted as phase turbulence. Conversely, the spatiotemporally disordered regimes of some systems believed to be fit for a phase-only description are often called phase turbulence regimes, even though it is often not made clear whether the phase equations which could be derived are well-behaved and/or correctly account for the chaotic dynamics of the original system.

A PARTICULAR BUT CRUCIAL CASE

We now turn to two intimately related equations which occupy a predominant place in phase turbulence problems, the Kuramoto-Sivashinsky and the complex Ginzburg-Landau equations.

The Kuramoto-Sivashinsky (KS) equation was derived independently by Kuramoto and Tsuzuki[6] to account for the dynamics of the temporal phase of the limit cycle emerging in some reaction-diffusion systems and by Sivashinsky[7] to describe the modulations of a propagating flame front (the "phase" is then just the position of the interface). It also appears naturally in other contexts, for example waves at the surface of a fluid running down an inclined plane. It reads, in

one space dimension[†]:

$$\partial_t \theta + \partial_{x^2}\theta + \partial_{x^4}\theta + (\partial_x\theta)^2 = 0 \qquad (2)$$

The large amount of literature on the KS equation published over the last fifteen years[8] is mostly due to the fact that it is probably the simplest partial differential equation exhibiting chaotic behavior. In most cases, it is indeed a phase equation and its spatiotemporally disordered regimes can thus legitimately be classified as phase turbulence (fig. 1), especially in the large-size limit where chaos becomes *extensive* —i.e. the amount of chaos, as measured e.g. by the number of positive Lyapunov exponents, is proportional to the system size[9].

At a general level, the KS equation is the phase equation of parity-invariant systems near a phase instability, when the phase diffusion coefficient D is small but negative. In this case, eq. (1) has to be supplemented by higher-order terms in order to possibly provide a phase equation in which the gradients do remain small. Symmetry considerations lead to a fourth-order equation which *a priori* reads:

$$\begin{aligned}
\partial_t\phi &= D\partial_{x^2}\phi - K\partial_{x^4}\phi \qquad (3)\\
&+ g_0(\partial_x\phi)^2 + g_1(\partial_x\phi)(\partial_{x^3}\phi)\\
&+ g_2(\partial_{x^2}\phi)^2 + g_3(\partial_x\phi)^2(\partial_{x^2}\phi)
\end{aligned}$$

To be more specific, and show how eq. (3) can be reduced to the KS equation, let us derive the phase equations associated to the complex Ginzburg-Landau (CGL) equation. Although it is taken here as the "reference system", it is actually the amplitude equation governing the slow modulations of an extended system having undergone a Hopf bifurcation ($k_0 = 0$, $\omega_0 \neq 0$)[1-4]. In one space dimension, the CGL equation reads:

$$\begin{aligned}
\partial_t A &= \varepsilon A + (1 + i\alpha)\partial_{x^2}A\\
&- (1 + i\beta)|A|^2A \qquad (4)
\end{aligned}$$

[†] In this formulation, where all coefficients have been rescaled to order 1, the gradients of θ are not small, in contrast to those of the original variable ϕ

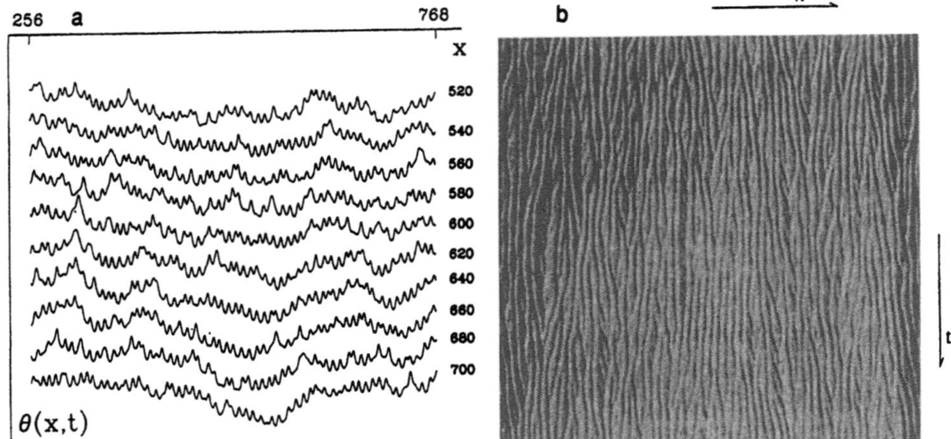

Figure 1: Phase turbulence in the KS equation: Results of a simulation using a pseudo-spectral code with periodic boundary conditions at a distance $L = 1024$; only the central part of length 512 is shown; low level initial conditions were used. The duration of the displayed sequence is $\Delta T = 256$; it starts after the elimination of a transient of length 500. (a) Flame-front like progression (time interval between two plots: $\Delta t = 20$). (b) Space-time grey-level plot of $\partial_x\phi$; 256 equally spaced levels from -4 to +4: $\partial_x\phi$ is not small here but eq(2) is a rescaled version of the "original" phase equation (see text).

where A is a complex amplitude, α and β two real parameters, and ε the relative distance to the (Hopf bifurcation) threshold. The "reference solution" U is now the spatially-homogeneous solution:

$$A_0 = \sqrt{\varepsilon}\exp(i\omega_0 t) \quad (5)$$
$$\text{with} \quad \omega_0 = -\beta\varepsilon$$

The slowly perturbed solution is written:

$$A = (\sqrt{\varepsilon} + \varrho)\exp i(\omega_0 t + \phi). \quad (6)$$

The gradient $\partial_t\phi$ and the central manifold are *a priori* expanded as[†]:

$$\partial_t\phi = \mathcal{F}(\phi, \partial_x) = u_1\partial_x\phi + u_2\partial_{x^2}\phi$$
$$+ u_3(\partial_x\phi)^2 + \cdots$$
$$\varrho = \mathcal{G}(\phi, \partial_x) = v_1\partial_x\phi + v_2\partial_{x^2}\phi$$
$$+ v_3(\partial_x\phi)^2 + \cdots \quad (7)$$

Injecting eqs. (5-7) in (4) and solving term by term at fourth-order, one obtains eq. (3)

with the following coefficients:

$$\begin{aligned}
D &= 1 + \alpha\beta \\
K &= \alpha^2(1 + \beta^2)/(2\varepsilon) \\
g_0 &= \beta - \alpha \\
g_1 &= 2g_2 = \alpha g_3 = -2\alpha(1 + \beta^2)/\varepsilon
\end{aligned} \quad (8)$$

Within the CGL equation, the phase instability thus appears whenever $D = 1 + \alpha\beta < 0$ (Newell's criterion). Eq. (3) with coefficients (8) reduces to the KS equation in the vicinity of the $1 + \alpha\beta = 0$ line of the parameter space (the so-called Benjamin-Feir line). Indeed, choosing D as a small parameter and imposing $\partial_x\phi \sim D^{3/2}$ and[†] $\phi \sim D$, all the extra nonlinear terms can be neglected and eq. (2) is recovered after proper rescaling of x, t and ϕ[4].

The KS equation thus correctly describes the phase dynamics of the CGL equation only *asymptotically close* to the Benjamin-Feir line $(1 + \alpha\beta \lesssim 0)$. Elsewhere in the phase-unstable region, the full phase

---

[†]A this stage, symmetry considerations can also be taken into account, simplifying the *a priori* expansion by retaining only relevant terms.

[†]This restriction on ϕ might appear "unphysical", as ϕ is a phase; in fact it merely amounts to choosing 0 as a convenient reference level for ϕ.

equation (3) has to be considered. Its dynamics is not well-behaved: infinitely-large phase gradients appear in finite time[13, 21]. In principle, this calls for a phase gradient expansion pushed to higher order(s). As a matter of fact, preliminary investigations indicate that these finite-time singularities are still observed then, and that the phase gradient expansion is riddled by convergence problems, putting into question the overall relevance of the phase dynamics approach in this case. Indeed, if one has to retain all terms in the expansion, no simplification is achieved, since the phase equation is then nothing but a complicated rewriting of the CGL equation. Whether this failure of the phase-only description is related to the dynamics exhibited in the phase-unstable region of parameter space by the CGL equation itself (i.e. other sources of fluctuations cannot be legitimately neglected) is, consequently, a question of interest.

Recent and ongoing work aims at establishing the "phase diagram" of the CGL equation in the large-size limit[10-12]. Numerical studies have confirmed the existence, in the phase-unstable region of the (α, β) plane, of spatiotemporally disordered regimes in which $|A|$ (resp. $\partial_x \phi$) remains bounded away from zero (resp. infinity). These regimes are themselves called, somewhat abusively in view of the above discussion, phase turbulence regimes. They are numerically observed in a region of parameter space which shrinks as the space dimension increases[21]. Whereas for $1 + \alpha\beta \lesssim 0$ they strongly resemble the chaotic regimes of the KS equation, other dynamical features play an important role at finite distance of the $1 + \alpha\beta = 0$ line (fig. 2).

The "breakdown of phase turbulence" in the CGL equation is signalled by the occurrence of a "defect" (a zero of A) preceded by a local divergence of the phase gradient. This scenario, consistent with the finite-time singularity observed in the phase equations, is easily observed numerically[12, 13]. It defines the borderline of the domain of existence of phase turbulence in the (α, β) plane.

Naturally, numerical results are only for large but finite systems observed during long but finite times. A crucial question is the status of phase turbulence in the infinite-size, infinite-time, "thermodynamic" limit. If, in this limit, the phase turbulence region shrinks and collapses on the $1 + \alpha\beta = 0$ line, then phase turbulence does not exist as a steady phenomenon not only in the phase approach, but also in the original equation, and these two facts are probably connected.

On the other hand, it might very well be that there exist, even in the thermodynamic limit, spatiotemporally disordered regimes in which no zeros of A ever occur, but that these regimes cannot be described by any finite-order phase equation even though the phase is always defined everywhere. Careful, intensive numerical simulations are under way in order to provide data on the status of phase turbulence in the thermodynamic limit by studying quantitatively finite-size effects[21]. At any rate, there is considerable evidence that even if it is ultimately a transient phenomenon, phase turbulence can be observed long enough so as to deserve attention *per se*.

As a matter of fact, other conjectures related to phase turbulence are currently being vividly debated, especially for the KS and CGL equations. At stake is the nature of the long-wavelength properties of phase turbulence and their mapping onto stochastic interface models. Indeed, both the KS equation and the CGL equation (in a phase turbulence regime) can be seen as deterministic interface models in which the position of the interface is simply $\phi(x, t)$ (note that this is exactly the meaning of ϕ in the "flame-front origin" of KS). It has been argued for some time[15-20] that the long-wavelength properties of such interfaces should be governed by an effective nonlinear Langevin-like equation now often called the Kardar-Parisi-Zhang (KPZ)

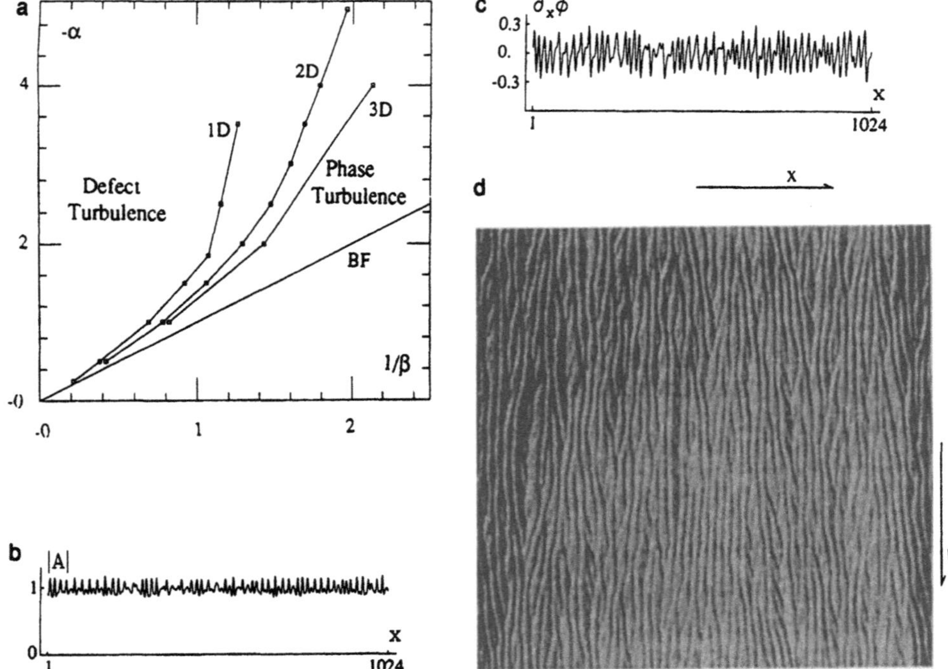

Figure 2: Phase turbulence in the CGL equation. (a) Phase turbulence domain in the $(1/\beta, -\alpha)$ parameter plane for $d = 1, 2, 3$. Typical aspects of $|A|$ (b) and $\partial_x\phi$ (c) for $1 + \alpha\beta = -1.5$, i.e. far enough from BF in the 1D case; notice the presence of propagative structures on the space-time grey-level plot of $|A|$ (d), which departs from the dynamics of the KS equation (see fig. 1b). (Results of a simulation using a pseudo-spectral code with periodic boundary conditions at a distance $L = 1024$).

model[14]:

$$\partial_t h = \nu \partial_{x^2} h + \lambda(\partial_x h)^2 + \eta(x, t)$$

where h is the coarse-grained position of the interface, $\nu > 0$ and λ are two real parameters, and η is a δ-correlated Gaussian noise.

Underlying this whole problem is the question of the "universality" of the KPZ model, whose scaling properties are exhibited by a large number of growth models[14]. Perhaps the two most crucial tests of the proposed relevance of the KPZ equation to describe the long-wavelength properties of phase turbulence in the KS and CGL equations are:

– the nature of the decay of the space autocorrelation function in two space dimensions (exponential, stretched exponential, or algebraic);

– the existence of long-range rotating order in three space dimensions.

In this context, phase turbulence appears to be one of the best playgrounds in which to confront ideas and theories trying to link spatiotemporally chaotic deterministic systems and modern nonequilibrium statistical mechanics.

Bibliography

• General reading on **phase dynamics** and phase equations can be found in the following books and reviews. The book by Y. Kuramoto and the dissertation by J. Lega focus particularly on phase turbulence and the CGL equation. For amplitude equations, see the corresponding article in this volume.

[1] Y. Kuramoto, *Chemical Oscillations, Waves and Turbulence*, Springer, Tokyo, 1984.

[2] P. Manneville, *Dissipative Structures and Weak Turbulence*, Academic Press, New-York, 1990.

[3] M.C. Cross and P.C. Hohenberg, "Pattern formation outside of equilibrium", Rev. Mod. Phys. **65** (1993) 851.

[4] J. Lega, "Défauts topologiques associés à la brisure de l'invariance de translation dans le temps," Thèse de doctorat, Université de Nice (1989).

[5] H. Brand, "Phase dynamics — a review and a perspective," in *Propagation in Systems Far From Equilibrium*, E.J. Wesfreid *et al.* eds. (Springer, 1987).

• Original articles where the **Kuramoto-Sivashinsky** equation has been derived are:

[6] Y. Kuramoto and T. Tsuzuki, "Reductive perturbation approach to chemical instabilities," Prog. Theor. Phys. **52** (1974) 1399.

[7] G.I. Sivashinsky, "Nonlinear analysis of hydrodynamic instability in laminar flames — I. Derivation of basic equations," Acta Astronautica **4** (1977) 1177; "Instabilities, pattern formation, and turbulence in flames," Ann. Rev. Fluid Mech. **15** (1983) 179.

• A review of results on the KS equation itself and can be found in:

[8] P. Manneville, "The Kuramoto-Sivashinsky equation: a progress report," in *Propagation in Systems Far From Equilibrium*, E.J. Wesfreid *et al.* eds. (Springer, 1987).

• The extensive chaos aspect of the KS equation was first exposed in:

[9] Y. Pomeau, A. Pumir, P. Pelcé, "Intrinsic stochasticity with many degrees of freedom," J. Stat. Phys. **37** (1984) 39; P. Manneville, "Lyapunov exponents for the Kuramoto-Sivashinsky model," in *Macroscopic modeling of turbulent flows*, O. Pironneau ed., Lecture Notes in Physics **230** (Springer, 1985).

• A very good presentation of the **complex Ginzburg-Landau equation** and its phase equations can be found in J. Lega's dissertation (see above). Recent work on the phase diagram and phase turbulence include:

[10] B.I. Shraiman, A. Pumir, W. van Saarloos, P.C. Hohenberg, H. Chaté and M. Holen, "Spatiotemporal chaos in the one-dimensional complex Ginzburg–Landau equation," Physica D **57** (1992) 241.

[11] H. Chaté, "Spatiotemporal intermittency regimes of the one-dimensional complex Ginzburg-Landau equation", Nonlinearity **7** (1994) 185.

[12] H. Chaté, "Disordered regimes of the one–dimensional Ginzburg–Landau equation," to appear in *Spatiotemporal patterns in non-equilibrium complex systems*, Santa Fe institute series in the sciences of complexity (Addison-Wesley, 1994).

[13] H. Sakaguchi, "Breakdown of the phase dynamics," Prog. Theor. Phys. **84** (1990) 792.

• The **Kardar-Parisi-Zhang equation** and its "universality" are reviewed in:

[14] J. Krug and H. Spohn, "Kinetic roughening of growing surfaces," in *Solids Far From Equilibrium*, C. Godrèche ed., (Cambridge University Press, 1991).

• Conjectures, theories, numerical work investigating the possible relationship between the KS, CGL and KPZ equations can be found in:

[15] V. Yakhot, "Large-scale properties of unstable systems governed by the Kuramoto–Sivashinsky equation," Phys. Rev. A **24** (1981) 642.

[16] S. Zaleski, "A stochastic model for the large scale dynamics of some fluctuating interfaces," Physica D **34** (1989) 427.

[17] K. Sneppen, J. Krug, M.H. Jensen, C. Jayaprakash, and T. Bohr, "Dynamic scaling and crossover analysis for the Kuramoto–Sivashinsky equation," Phys. Rev. A **46** (1992) 7351.

[18] I. Procaccia, M.H. Jensen, V.S. L'vov, K. Sneppen, and R. Zeitak, "Surface roughening and the long-wavelength properties of the Kuramoto–Sivashinsky equation," Phys. Rev. A **46** (1992) 3220; V.S. L'vov and I. Procaccia, "Comparison of the scale invariant solutions of the Kuramoto–Sivashinsky and Kardar–Parisi–Zhang equations in d dimensions," Phys. Rev. Lett. **69** (1992) 3543

[19] C. Jayaprakash, F. Hayot, and R. Pandit, "Universal properties of the two-dimensional Kuramoto–Sivashinsky equation," Phys. Rev. Lett. **71** (1993) 12.

[20] G. Grinstein, C. Jayaprakash, and R. Pandit, "Conjectures about phase turbulence in the complex Ginzburg–Landau equation," preprint 1993.

[21] H.Chaté and P. Manneville, to be published.

PREDICTABILITY
(in Turbulence)

G. PALADIN
Dipartimento di Fisica, Università dell'Aquila,
Via Vetoio, I-67010 Coppito L'Aquila, Italy
M.H. JENSEN
The Niels Bohr Institute,
Blegdamsvej 17, DK-2100 Copenhagen Ø, Denmark
A. VULPIANI
Dipartimento di Fisica,
Università "La Sapienza",
P.le Aldo Moro 2, I-00185 Roma, Italy

The concept of predictability plays a major role in the theory of chaotic dynamical systems and turbulence. In dynamical systems, the predictability is related to the sensitive dependence on initial conditions, i.e. to the fact that the distance of two nearby orbits diverges as $\exp(\lambda t)$ where λ is the maximum Lyapunov exponent[1]. Consequently if the maximum admitted error of the state of the system is δ_{max} and the initial error is δ_0, then the future of the system can be predicted up to a time

$$T_p \sim \frac{1}{\lambda} \ln \frac{\delta_{max}}{\delta_0} \qquad (1)$$

This simple remark is rather important since it implies that in dynamical systems the forecasting is limited by the chaotic nature of the evolution and not by the resolution of the measurements. The gain obtained by achieving finer resolutions is only logarithmic and can be safely ignored for practical purposes.

The problem of predictability in low dimensional systems is thus solved in the context of the theory of chaotic dynamics by considering of the evolution of a perturbation on the original flow. In fully developed turbulence the predictability time is less well defined because one deals with partial differential equations. A classical argument of Landau shows that the degrees of freedom necessary to get a good description of a three dimensional turbulent fluid increases as a power of the Reynolds number $Re^{9/4}$. In dynamical systems with a large number of degrees of freedom, one should consider not only the rate λ of divergence of nearby trajectories, but also the direction in the phase space where a perturbation grows. The predictability on those systems has another highly non-trivial feature: the propagation in space of a perturbation might be unrelated to the chaoticity degree of the global dynamical system. This point can be easily understood and analyzed in the context of coupled discrete maps on a lattice. If the interactions among the maps is local, the propagation of a perturbation is obviously independent of the Lyapunov exponent. One expects that it propagates from one lattice point toward the boundaries at the sound speed. On the other hand, for non-local interactions the situation is much less clear and should discussed case by case. For instance in one dimensional coupled maps, one observes that if the decay of interaction is slow enough (e.g. is proportional to the inverse of the distance from the perturbed point) the predictability time is given by (1). In turbulence, one can think that non-local interactions among different points of the fluid

Turbulence: A Tentative Dictionary
Edited by P. Tabeling and O. Cardoso, Plenum Press, New York, 1995

arise as a consequence of the incompressibility condition so that the problem of relation between Lyapunov and predictability is rather delicate.

The analytic calculation of the maximum Lyapunov exponent λ or of the predictability time is extremely difficult. However, it is possible to use phenomenological arguments to show that λ increases as a power of Reynolds. As a consequence, the predictability time could vanish with Reynolds if the interactions are sufficiently non-local, or in the opposite case it could be independent of the Reynolds number and related to the characteristic life-time of the largest eddies.

To determine the dependence on the Reynolds number, Ruelle pointed out that the maximum Lyapunov exponent should be proportional to the smallest characteristic time of the system[2]. It is the turnover time of an eddy of size comparable to the Kolmogorov viscous cut-off η. By dimensional counting, the turnover time of an eddy of size ℓ is $\tau(\ell) \sim \ell^{1-h}$, where h is the scaling exponent of the velocity difference $|u(x+\ell) - u(x)| \sim \ell^h$. The viscous cut-off vanishes as a power of the Reynolds number Re, i.e. $\eta \sim Re^{-1/(1+h)}$. These dimensional relations imply that the maximum Lyapunov exponent should scale as

$$\lambda \sim \tau(\eta)^{-1} \sim Re^\alpha \quad with \ \alpha = \frac{1-h}{1+h}.$$

In the Kolmogorov theory $h = 1/3$ for all space points α, so that $\alpha = 1/2$.

However, one expects that the presence of quiescent quasi-laminar periods should change the chaotic features of the fluid flow. In fact, the intermittency of energy dissipation can be described by introducing a spectrum of scaling exponents h. In the multifractal approach, the probability that the velocity difference $|u(x+\ell) - u(x)|$ scales as $\sim \ell^h$ is assumed to be $P_\ell(h) \sim \ell^{3-D(h)}$, where the function $D(h)$ is given by the Legendre transform of the exponent for the structure functions[3]. The multifractality also implies the existence of a spectrum of viscous cut-offs, since each h selects a different damping

scale $\eta(h) \sim Re^{-1/(1+h)}$, and hence a spectrum of turnover times. To find the Lyapunov exponent, one should integrate over the h-distribution

$$
\begin{aligned}
\lambda \ &\sim \ \int \tau(h)^{-1} P_\eta(h) \, dh \\
&\sim \ \int \eta^{h-D(h)+2} \, dh
\end{aligned}
\tag{2}
$$

where $\tau(h)$ is the turnover time of an eddy of scale $\eta(h)$, so that $\tau(h) \sim \eta(h)^{1-h} \sim Re^{-(1-h)/(1+h)}$. For large Re the viscous cut-offs $\eta(h)$ is very small and (2) can be estimated by the saddle point method,

$$\lambda \sim Re^\alpha \ with \ \alpha = \max_h \frac{D(h) - 2 - h}{1+h} \tag{3}$$

The value of α depends on $D(h)$. By using the function $D(h)$ obtained with the random beta model fit[3, 4], one gets $\alpha = 0.459..$, smaller than the Ruelle prediction $\alpha = 0.5$[2].

In order to characterize the finite-time fluctuation of the degree of chaos one can introduce an effective Lyapunov exponent γ_τ[3], defined as the exponential divergence rate after a time delay τ of two trajectories close at time t, i.e.

$$\gamma_\tau(t) = \frac{1}{\tau} \ln \frac{||\delta u(t+\tau)||}{||\delta u(t)||} \tag{4}$$

where δu is the infinitesimal difference between two velocity fields evolving under the same equations, i.e. the $3D$ Navier-Stokes equations.

The Lyapunov exponent is given by $\lambda = \lim_{\tau \to \infty} \gamma_\tau$ which has the same value for almost all initial conditions. However, there are finite τ fluctuations and, in general cases, the probability of finding $\gamma_\tau \neq \lambda$ scale as $P(\gamma) \sim \exp[-S(\gamma)\tau]$, where $S(\gamma) \geq 0$ with the equal sign for $\gamma = \lambda$. When the response to a perturbation after a time τ factorizes into a product of uncorrelated variables, central limit arguments can be applied so that $S(\gamma) \sim (\gamma - \lambda)^2/\mu$, where $\mu = \lim_{\tau \to \infty} \tau \langle (\gamma_\tau - \lambda)^2 \rangle$. In general, the parabolic shape of $\cdot S$ remains valid for small $|\gamma - \lambda|$. The parameters λ and μ give the main characterization of the γ-distribution since it is possible to show that the dynamical intermittency becomes relevant when $\mu/\lambda > 1$ (for a detailed discussion of this point see Ref. [3]).

These arguments can be tested in the shell model of Ohkitani and Yamada (O-Y model, see the related contribution in this book)[5, 6] which exhibits a multifractal scaling with a function $D(h)$ similar to that found in experimental data[7]. In the O-Y model, multifractality is due to the intermittency in the dynamical evolution, since energy bursts are observed to interrupt quiescent laminar periods when there is a sudden increase of the effective Lyapunov exponent. In fact, the degree μ/λ of intermittency diverges with Reynolds, since a numerical calculation[3, 4] gives $\mu \sim Re^{0.8}$ while $\lambda \sim Re^{0.46}$. As a consequence, there is a transition from weak to strong intermittency at $Re \approx 10^6$ where $\mu/\lambda \approx 1$. The existence of two different dynamical regimes has important reflexes on the statistics of the predictability time T. Here, T is defined as the time needed for a disturbance δ on the velocity field, localized at the Kolmogorov scale $\eta(h = 1/3)$, to affect the large scales. In practice, one looks at the time necessary for a small perturbation of the velocity field on the dissipative shell to become larger than a given threshold value g in one of the first shells. Numerically, the time T is strongly dependent on the degree of chaos. If the system undergoes a burst of energy (and of effective Lyapunov exponent), the predictability time is very small, while if the system is in a laminar period, the predictability time is very large. It follows that the probability distribution function (PDF) of T has very different shapes depending on whether μ/λ is less or larger than unity. Up to $Re \approx 10^6$, one observes a rather peaked PDF with an almost Gaussian form. At larger values of Re, the distribution gets a long exponential tail, which indicates the possibility of large excursions in the value of T, depending on

whether the system is in a turbulent or in a purely laminar period. Furthermore, the typical predictability time T_t (given by the maximum of the PDF) is fully related to the Lyapunov exponent. In the O-Y shell model, one roughly has $T_t \sim 1/\lambda$, so that the typical predictability time decreases as a power of Re. Moreover, at increasing Re the occurrence of large values of $(T - T_t)/T_t$ is more and more likely[4, 8].

These results are in contrast with the conjecture that the predictability time is proportional to the life-time of the energy containing eddies, and hence independent of Re[9]. This "classical" result about predictability in three dimensional turbulence is based on physical arguments and closure approximations. The approach is the following. A perturbation at wavenumber $2k$ induces a complete uncertainty on the velocity field on the wave number k after a time $\tau(k)$, assumed to be proportional to the typical eddy turnover time at scale k:

$$\tau(k) \sim \frac{1}{k u_k}$$

where u_k is the typical velocity difference at scale $1/k$. In the Kolmogorov theory $\tau(k) \sim k^{-2/3}$ and the predictability time for an uncertainty to propagate from the Kolmogorov scale $\eta \sim k_K^{-1}$ up to the scale of the energy containing eddies $L_0 \sim k_0^{-1}$, is given by:

$$T \simeq \sum_{n=0}^{N} \tau(2^n k_K) \sim L_0/v_0 \qquad (5)$$

where $N = \ln_2(k_K/k_0) \sim \ln Re$ and $k_K \sim Re^{3/4} k_0$. Closure approximations[9], where one still uses dimensional arguments, give the same results. In the above arguments many characteristic times are involved so that (5) strongly depends on the physical mechanism for the inverse cascade of the perturbation. Note that (5) is not sensitive to the presence of intermittency.

On the contrary in the O-Y shell model one has[4, 8] $T \sim Re^{-0.46}$, and hence a strong dependence on the Reynolds number. This result involves only one characteristic time, the eddy turnover time at the

Kolmogorov scale. On physical grounds, it could seem rather strange that the predictability on large scale is related to properties at the Kolmogorov scale. Unfortunately, it is not easy to reach a clear understanding of this problem which is strictly related to the relevance of non-local interactions among different spatial points.

In obtaining (5), where T is independent of the Reynolds number and of intermittency, the locality of interactions plays a central role and one obtains features which are similar to those of a system of coupled maps with only local couplings. On the contrary, the O-Y model has predictability properties analogous to those of non-local coupled maps.

The intermittency of the O-Y shell model has another peculiar feature which has relevant physical consequences on the growth of a disturbance field: the energy bursts interrupt quiescent laminar periods when there is a sudden increase of chaoticity localized along directions in the phase space which correspond to the small length scales, i.e. to the last shells close to the viscous cut-off[8].

To study the "butterfly effect" one should introduce a small perturbation ϵ at high wavenumbers, close to the dissipative cut-off (the Kolmogorov length). Let us denote the state of the system at time t as $u(t) \equiv \{ u_1(t), \cdots, u_n(t) \}$, where u_n is the typical velocity difference at scale $k_n^{-1} \sim 2^{-n}$. A perturbed state $u'(t)$ is introduced at time t by adding a small increment ϵ to the velocity component in some of the shells. Next, define the distance between two initially nearby trajectories as $D(\tau) = |u(t+\tau) - u'(t+\tau)|$. In chaotic systems, it grows exponentially, i.e. $< \ln D(\tau) > \approx \lambda \tau$.

Consider the difference at the nth shell at time τ is $\delta_n(\tau) = |u_n(t+\tau) - u'_n(t+\tau)|$. As a consequence of the particular form of the dynamical intermittency of the O-Y shell model, the exponential growth of $\delta_n(\tau)$ is triggered by a large intermittent burst localized at the small length scales and associated with a sharp increase of the effective Lyapunov exponent γ defined in (4). After such a burst, the value of $\delta_n(\tau)$ increases with τ at smaller and smaller k_n, so that the initially localized disturbance propagates towards lower k_n by a kind of inverse cascade. Finally, the disturbance reaches the beginning of the inertial range and the flow at large scales is affected: the "butterfly" disturbance has grown to macroscopic scales.

It is also interesting to make a perturbation at small wavenumbers k_n (large length scales) and monitor how the disturbance grows. Surprisingly, there is no exponential growth at any of the k_n-values before a component of the disturbance has spread all the way to large k_n. The time needed by this "precursor" disturbance to move from small to large wavenumbers is very fast, of the order λ^{-1}. After this, the exponential growth is again triggered by a chaotic burst localized on large k_n, leading to a sort of inverse cascade of the instability. In the O-Y shell model, the butterfly effect always stems from the small length scales close to the dissipative cut-off even if the perturbation is performed at large scales.

In conclusion, the multifractal description of three dimensional turbulence is relevant for a deeper understanding of predictability. In particular, the scaling of the Lyapunov exponent with the Reynolds number can be expressed in terms of the multifractal spectrum $D(h)$ and multifractality gives rise to a strongly intermittent chaotic regime at high Reynolds number. In fact, the variance of the fluctuations of the effective Lyapunov exponent is found to diverge with the Reynolds number.

We have also argued that the Lyapunov exponent is related to the predictability time. This is a crucial and controversial point. Such a relation holds in the O-Y shell model, which allows one to formulate two conjectures which need an experimental verification in three dimensional turbulent fluids:

(1) the occurrence of long predictability times becomes more probable at

increasing Re. In this sense, fully developed turbulence could exhibit a smooth transition from a quasi-gaussian toward a strongly intermittent probability distribution of the predictability times (from "weak" toward "strong" turbulence);

(2) efficient exponential growth of a perturbation should start only from small length scales and only during an energy burst.

Bibliography

[1] G. Benettin, L. Galgani and J.M. Strelcyn, Phys. Rev. A **14**, 2338 (1976).

[2] D. Ruelle, Phys. Lett. **72A**, 81 (1979). D. Ruelle, "Hasard et chaos" (Editions Odile Jacob, Paris 1991).

[3] G. Paladin and A. Vulpiani, Phys. Rep. **156**, 147 (1987).

[4] A. Crisanti, M.H. Jensen, G. Paladin and A. Vulpiani, Phys. Rev. Lett., **70**, 166 (1993).

[5] M. Yamada and K. Okhitani, J. Phys. Soc. of Japan **56**, 4210 (1987); M. Yamada and K. Okhitani, Progr. Theo. Phys. **79**, 1265 (1988); M. Yamada and K. Okhitani, Phys. Rev. Lett. **60**, 983 (1988).

[6] M.H. Jensen, G. Paladin and A. Vulpiani, Phys.Rev.A **43**, 798 (1991).

[7] F. Anselmet, Y. Gagne, E. J. Hopfinger and R. Antonia, J. Fluid Mech. **140**, 63 (1984).

[8] A. Crisanti, M.H. Jensen, G. Paladin and A. Vulpiani, J. Phys. **A**, in press (1993).

[9] C.E. Leith and R.H. Kraichnan, J. Atmos. Sci. **29**, 1041 (1972). D.K. Lilly, in "Dynamic Meteorology" pag. 353 Ed. P. Morel (D. Reidel Publishing Company, Boston 1973).

PROBABILITY DENSITY FUNCTIONS
(in 3D Turbulence)

B. CASTAING

Centre de Recherches sur les Très Basses Températures,
Laboratoire associé à l'Université Joseph Fourier,
BP 166, F-38042 Grenoble-cedex 9, France

INTRODUCTION

For a long time, fluid mechanicists have observed that the probability density of various quantities (scalar density, velocity differences, ...) in turbulent flows strongly differ from the gaussian shape one could naively expect, due to the random character of turbulence[1, 2]. Only recently however the problem of the shape of some of these probability density functions (pdf) has been addressed[3-7]. The attempts concerned the incompressible flows and two different kinds of pdfs:

i) The pdf of difference of some quantities (velocity, temperature, ...) between two points, the distance between these points belonging to the inertial range. Since Kolmogorov's papers[1] this range, which corresponds to the scales small compared to the energy containing ones, is expected to display universal properties.

ii) The single point pdf of some quantities like the pressure or the temperature (acting as passive or active scalar). According to the generally accepted ideas, these pdfs should depend on the boundary conditions or the type of flow. The question then turns around the existence of generic boundary conditions, determining the shape of the pdf, as the existence of a large scale scalar density gradient for instance[8].

Very different processes have been used to derive the shape of the pdfs. Some of them could be generalized:

- The most natural assumption for a stationary process is a "stability principle": the future pdf obtained from a model for the flow evolution must be identical to the present one. This principle has been for instance used for the one point passive scalar pdf[9], and in some way for the two points (difference)[10] pdfs.

- In a more questionable way, an "ergodicity principle" has been invoked, the one point stationary pdf being identified with the instantaneous pdf of the whole flow. This has been used for the pressure pdf[11] and the temperature one[12, 13] in Rayleigh-Bénard convection.

In view of all these approaches, giving different results, it is clear that a precise characterization of the obtained pdf shape is essential for discriminating between the theories. This is, generally speaking, a weak point of most of the studies. Many of them use in a loose way the term "exponential" for characterizing the wing of the pdf. Some others compare the pdf to a one parameter family of functions (stretched exponential, ...), with the hope that the "shape parameter" could be predicted by a theory. Very few are completing this program. Separate predictions for the positive and negative sides of the pdf suffer from a great weakness: they exclude the center, in general without precise definition of the extension of this center. A comparison with experiment is then hardly decisive.

THE DIFFERENCE PDF - INTERMITTENCY

While the ideas which shall be exposed could be generalized at least to the scalar concentration differences, the most studied case concerns the velocity difference between two points[2-4,14-26]. The interest in their pdfs has been renewed by experimental[3,14-17] and numerical[4,18-20] studies, but also by models[21-24], mainly devoted to the gradient pdf. The goal, in the last approaches, is to grasp the minimum physical input which yields the observed shapes. Without disregarding the interest of these works, we shall develop a more analytic approach, aiming to extract comprehensive quantities from the pdf shapes and their evolution.

Experimentally the easiest components of this velocity difference to study is the longitudinal one, parallel to the vector $\vec{r}$ joining the two points:

$$\delta v = v_r(\vec{x} + \vec{r}) - v_r(\vec{x})$$

An exact result, within the hypothesis of isotropy has been obtained by Kolmogorov[1] for these fluctuations:

$$\begin{aligned}
<\delta v^3> &= -\frac{4}{5}<\epsilon^*>r \\
&\quad +6\nu\frac{\partial}{\partial r}<\delta v^2> \quad (1)
\end{aligned}$$

where $<>$ stands for ensemble average and $<\epsilon^*>$ is the average dissipation per unit mass. Then Kolmogorov postulated that the statistics of δv in the inertial range is fully controlled by $<\epsilon^*>$ which dimensionally yields:

$$<\delta v^n> = C_n <\epsilon^*>^{n/3} r^{n/3}, \quad (2)$$

equivalent to a universal distribution for the reduced variable $\delta v/\sigma$ with $\sigma = <\epsilon^*>^{1/3} r^{1/3}$.

This is clearly wrong. The experiment shows an evolution of the shape of δv's pdf when r goes from the large scales to the small ones. This has been taken into account by a distribution of scaling factors σ at each scale r, instead of the single value $<\epsilon^*>^{1/3} r^{1/3}$. Writing the pdf of δv as:

$$\frac{1}{\sigma_r} P_r\left(\frac{\delta v}{\sigma_r}\right), \quad (3)$$

where $\sigma_r^2 = <\delta v^2>$, we have:

$$\frac{1}{\sigma_2} P_r\left(\frac{\delta v}{\sigma_r}\right) =$$

$$\int G_r(\ln \sigma) \frac{1}{\sigma} P_L\left(\frac{\delta v}{\sigma}\right) d\ln \sigma, \quad (4)$$

where L is the large (integral) scale. This formula got recently enhanced concrete and physical meaning due to the experimental results we relate now[17]. Longitudinal velocity differences are often measured with only one probe invoking Taylor's "frozen turbulence" hypothesis: the velocity fluctuations are assumed to be convected at the average velocity $\vec{U}$ which reads:

$$\vec{v}(\vec{x} + \vec{U}\tau, t) \simeq \vec{v}(\vec{x}, t - \tau)$$

and

$$\begin{aligned}
\delta v &= v_u(\vec{x} + \vec{U}\tau, t) - v_u(\vec{x}, t) \\
&\simeq v_u(\vec{x}, t - \tau) - v_u(\vec{x}, t)
\end{aligned}$$

On the same interval $[t - \tau, t]$ the following quantity can be calculated:

$$\epsilon = \frac{1}{\tau}\int_{t-\tau}^{t}\left(\frac{\partial v}{\partial t}\right)^2 d\tau - \left(\frac{\delta v}{\tau}\right)^2. \quad (5)$$

Invoking again the Taylor hypothesis, the first term gives an estimation of the dissipated power on the considered interval. Then the second term corresponds to the power dissipated at the scale $U\tau$ and ϵ could be seen as proportional to the energy transfer rate towards the smaller scales.

Conditioning the δv sampling to one particular value of ϵ, Gagne et al.[17] found that the shape of obtained conditional pdfs is universal, independent of ϵ or the Reynolds number, identical to the P_L shape. Thus, fixing ϵ is equivalent to fixing σ in the language of Eq.(4) and the physical idea of Kolmogorov and Obukhov[1, 27] attributing the intermittency to the distribution of the energy transfer rate is in some way confirmed.

Eq.(4) is compatible with most of the treatments of intermittency. In the Kolmogorov-Obukhov[27] approach:

$$G_r(\ln \sigma) = \frac{1}{\lambda\sqrt{2}} \exp -\frac{\ln^2 \sigma/\sigma_m}{2\lambda^2}, \quad (6)$$

i.e. a log-normal distribution of σ. In the multifractal frame[27], σ is assumed to locally depend on r as a power law:

$$\sigma = \sigma_L \left(\frac{r}{L}\right)^h, \qquad (7)$$

with a fractal distribution of the (singularity) exponents h.

The shape of G_r is an interesting question, on which the different models disagree. It is difficult to address experimentally however, because the shape of P_r poorly depends on the shape of G_r[28]. It mostly depends on the width $< (\delta \ln \sigma)^2 >$ which is thus the easiest parameter to get from experiments, for instance by using the log-normal ansatz (6) and fitting the experimental pdf via Eq.(5)[15]. The best value of λ^2 gives a good estimate for $< (\delta \ln \sigma)^2 >$[28].

The Kolmogorov-Obukhov[27] approach and the multifractal[16,28,30] one asymptotically predict a logarithmic dependence for $< (\delta \ln \sigma)^2 >$ versus r:

$$< (\delta \ln \sigma)^2 > = \mu \ln \frac{L}{r}, \qquad (8)$$

The variational approach[15] predicts a power law dependence for the same quantity:

$$< (\delta \ln \sigma)^2 > = \lambda_0^2 \left(\frac{r}{\eta}\right)^{-\beta}, \qquad (9)$$

where η is the small (dissipative) scale, with a logarithmic dependence of β versus the Taylor scale based Reynolds number R_λ[16]:

$$\beta = \frac{\beta_0}{\ln \frac{R_\lambda}{R^*}} \qquad (10)$$

Asymptotically, for $R_\lambda \to \infty$, Eq.(9) is equivalent to Eq.(8). However the effective μ is not the same in the neighborhood of $r \simeq \eta$ ($\mu = \beta \lambda_0^2$) and close to $r \simeq L$. Experimentally, for the finite Reynolds experimentally accessible, Eq.(9) seems to be valid on the whole inertial range. It has the important consequence that looking at the large inertial scales gives access to the Reynolds number, through the measure of β in Eq.(10).

THE ONE POINT PDF

Again the interest came here as well from experimental[3] as numerical[6, 7] studies.

The most significant progress concerns the passive scalar pdf[8, 9]. As noted in the introduction, one point pdfs are sensitive to the boundary conditions and an important point has been to choose expected generic ones, namely the presence of a large scale concentration gradient for the scalar. This gradient has to be uniform on a much larger scale than the turbulent integral scale and experiments have been recently performed under these conditions[31, 32]. Then it is argued that the fluid element at the origin at time $t = 0$ was at different place a time t before. During the path, fluctuations progressively wash down, but the element carries its average concentration up to a memory time t^*. The distribution of these average concentrations must reproduce the distribution of possible concentrations at the origin.

This distribution thus comes both from the various places one can reach in a time t^* and from the distribution of t^*. With a diffusion approximation for the fluid element behavior, the distribution of abscissa x reached within t^* is:

$$\pi(x) \propto \frac{1}{t^{*1/2}} \exp - \frac{x^2}{4Dt^*}.$$

The average concentration at x is $c(x) = c(0) + gx$. The distribution of concentrations is then :

$$P(\delta c) = \int G(t^*) \frac{1}{t^{*1/2}} \exp - \frac{\delta c^2}{4Dg^2 t^*} dt^* \qquad (11)$$

This looks like Eq.(10) which explains why the obtained shapes are similar ("exponential" or "stretched exponential"). Obviously the difficulty lies in a clean definition of t^* and in the estimation of its distribution $G(t^*)$, which could be performed only with a rather unrealistic model of turbulence[9]. But the method and its basic idea are promising.

For the active scalar, like the temperature in a Rayleigh-Bénard cell, this approach is in competition with another idea, that the distribution of temperature at a central point reflects the distribution of temperature in the whole cell[12, 13] (a kind of ergodic assumption for the fluid elements). In numerical simulations and experiments these two distributions seem to agree[13, 33]. Indeed, in the previous case the distribution of scalar concentration in the whole cell was much larger than the single point pdf. In such a case, a mean gradient approximation is always possible. When the two distributions are close in width, the validity of the "ergodic" principle remains to be confirmed.

With pressure fluctuations we find again the "ergodic" principle. The distribution of pressure at a wall in a von Kármán flow[11, 34] has been found close to the distribution of pressure in the whole field of a direct numerical simulation[7]. Theories relate this distribution to the gaussian distribution of velocities[35], or to the geometry of dissipative or vortical structures[11]. Both are in agreement with the "exponential-like" shape of the pdf for negative pressure fluctuations. But the positive wing seems gaussian[7] to a good approximation in disagreement with both theories.

OTHER APPROACHES

Another general approach has been initiated by Sinaï and Yakhot[36], and used by Ching[37]. The starting point is a general property of stationary processes, which allows the expression of the pdf in terms of two other functions of the random variable, normalized to its mean square deviation[38]: $< X >= 0, < X^2 >= 1$:

$$P(X) = \frac{1}{q(X)} \exp - \int_0^X \frac{r(x)dx}{q(x)}, \quad (12)$$

with

$$r(x) = \frac{< \ddot{X}|x >}{< \dot{X}^2 >}, q(x) = \frac{< \dot{X}^2|x >}{< \dot{X}^2 >}$$

$\dot{X}$ is the time derivative of X and $< A|B >$ is the ensemble average of A conditioned to the considered value of B.

We now have two unknown functions to find, r and q. But simple assumptions about these functions yield reasonable shapes for the pdf[37, 39]. In a complementary point of view, more information about the X statistics is contained in these two functions than in the pdf alone. Modeling them and understanding their shape is thus a new challenge.

CONCLUSION

The study of pdf's shapes, single point or difference ones, has the advantage to allow a direct comparison between experiments and numerical studies or simple modeling like the shell models[39]. We expect great progress in understanding developed turbulence from these comparisons in the next future. However, we have to keep in mind that it represents only one aspect of the turbulence and that many other ones have to be explored.

Bibliography

[1] A.S. Monin and A.M. Yaglom, in: Statistical Fluid Mechanics, vol. **2** (MIT Press, 1975).

[2] C.W. van Atta and J. Park, in: Statistical Models and Turbulence, Lecture Notes in Physics, vol. **12** (Springer, 1972) p. 402-426.

[3] Y. Gagne, in: Advances in Turbulence, vol. **3** (Springer, 1991).

[4] A. Vincent and M. Meneguzzi, J. Fluid Mech. **255** (1991) 1.

[5] F. Heslot, B. Castaing, and A. Libchaber, Phys. Rev. A **36** (1987) 5870.

[6] O. Métais and J. Herring, J. Fluid Mech. **202** (1989) 117.

[7] O. Métais and M. Lesieur, J. Fluid Mech. **239** (1992) 157.

[8] A. Pumir, B.I. Shraiman, and E.D. Siggia, Phys. Rev. Lett. **66** (1991) 2984.

[9] B.I. Shraiman, E.D. Siggia, to be published.

[10] B. Castaing, Y. Gagne, in: Turbulence in Spatially Extended Systems (ed. R. Benzi, Les Houches, 1993).

[11] S. Fauve, C. Laroche, B. Castaing, J. Phys. II (France) **3** (1993) 271.

[12] V. Yakhot, Phys. Rev. Lett. **63** (1989) 1965.

[13] R. Benzi, S. Succi, Europhys. Lett. **21** (1993) 305.

[14] F. Anselmet, Y. Gagne, E.J. Hopfinger and R.A. Antonia, J. Fluid Mech. **140** (1984) 63.

[15] B. Castaing, Y. Gagne and E. Hopfinger, Physica D **46** (1990) 177.

[16] B. Castaing, Y. Gagne, M. Marchand, Physica D **68** (1993) 387.

[17] Y. Gagne, M. Marchand, B. Castaing, to be published in J. Phys. II (brief comments).

[18] K. Yamamoto and I. Hosokawa, J. Phys. Soc. Jpn **57** (1988) 1532.

[19] S. Kida and Y. Murakami, Fluid Dyn. Res. **4** (1989) 347.

[20] Z.S. She, E. Jackson and S.A. Orszag, in: Proc. Newport Conf. on Turbulence (Springer, 1991).

[21] R. Benzi, L. Biferale, G. Paladin, A. Vulpiani and M. Vergassola, Phys. Rev. Lett. **67** (1991) 2299.

[22] J. Eggers and S. Grossmann, Phys. Rev. A **45** (1992) 2360.

[23] R.H. Kraichnan, Phys. Rev. Lett. **65** (1990) 575.

[24] Z.S. She, Phys. Rev. Lett. **66** (1991) 600.

[25] Z.S. She and S.A. Orszag, Phys. Rev. Lett. **66** (1991) 1701.

[26] Z.S. She, E. Jackson and S.A. Orszag, Proc. R. Soc. London A **434** (1991) 101.

[27] A.N. Kolmogorov, J. Fluid Mech. **13** (1962) 82; A.M. Obukhov, J. Fluid Mech. **13** (1962) 77.

[28] G. Parisi and U. Frisch, in: Turbulence and predictability in geophysical fluid dynamics, Proc. Int. School of Physics "E. Fermi" (Varenna, Italy, 1983) eds. M. Ghil, R. Benzi and G. Parisi (Amsterdam, North- Holland, 1985) p. 84.

[29] A. Naert, L. Puech, B. Chabaud, J. Peinke, B. Castaing, B. Hébral, J. Phys. (Paris) II 4, (1994) 215.

[30] C. Meneveau and K.R. Sreenivasan, Nucl. Phys. **B2** Proc. Suppl. (1987) 49.

[31] J.P. Gollub, J. Clarke, M. Gharib, B. Lane, and O.N. Mesquita, Phys. Rev. Lett. **67** (1991) 3507.

[32] Jayesh and Z. Warhaft, Phys. Rev. Lett. **67** (1991) 3503; Jayesh and Z. Warhaft, Phys. Fluids A **4** (1992) 2292.

[33] F. Chillá, S. Ciliberto, C. Innocenti, Europhys. Lett. **22** (1993) 681.

[34] Y. Couder, S. Douady, Private communication.

[35] M. Holzer, E.D. Siggia, Phys. Fluids A **5** (1993) 2525.

[36] Y.G. Sinaï and V. Yakhot, Phys. Rev. Lett. **63** (1989) 1962.

[37] E.S.C. Ching, Phys. Rev. Lett. **70** (1993) 283.

[38] S.B. Pope, E.S.C. Ching, Phys. Fluids A **5** (1993) 1529.

[39] L. Biferale, M.H. Jensen, G. Paladin, A. Vulpiani, Physica A **185** (1992) 19.

Rayleigh-Bénard Turbulent Convection

A. Tilgner, A. Belmonte, A. Libchaber[a]
Princeton University, Physics Department,
Jadwin Hall, Princeton, NJ 08544, USA
[a] also at NEC Research Institute,
4 Independence Way, Princeton, NJ 08540, USA

Rayleigh-Bénard convection is the flow of a fluid enclosed in a cell with bottom and top boundaries kept at fixed temperatures, the bottom being warmer than the top. Since the fluid in the vicinity of the bottom plate is warmer and therefore less dense than fluid in the upper part of the cell, an unstable layering ensues. Laboratory studies of convection have been chiefly confined to situations in which the "Boussinesq approximation" is justified. This is an approximation to the full equations of motion, which retains the temperature dependence of the density only in the buoyancy term and considers all other material properties to be uniform throughout the cell.

Three non dimensional numbers fully determine the convective flow in this case: 1) The aspect ratio, defined as the width of the cell divided by its height. 2) The Prandtl number Pr, defined as $Pr = \nu/\kappa$, where ν and κ are the kinematic viscosity and thermal diffusivity, respectively. Pr prescribes the relative importance of the two dissipative mechanisms opposing the flow. 3) The Rayleigh number Ra, given by

$$Ra = \frac{g\alpha\Delta L^3}{\kappa\nu}$$

where g is the earth's acceleration, α the thermal expansion coefficient, Δ the temperature difference applied to the cell, and L the height of the cell. Since Ra contains Δ, it specifies the driving of the turbulence. Convection starts at $Ra \approx 10^3$, where the driving is strong enough to overcome dissipation.

Just above onset, the flow therefore organizes itself into rolls or cellular patterns, which are about as wide as their height L. Thus only one roll fits into a cell of aspect ratio 1. This circulating flow, laminar at low Ra, exists also in the turbulent regimes[1, 2], in which of course large velocity fluctuations are superimposed on the mean flow. In cells of aspect ratio greater than one, in which several rolls exist near onset, the turbulence appears to destroy the rolls in favor of a single circulation following the perimeter of the cell.

The variation of mean temperature measured as a function of the distance of one of the temperature regulated plates exhibits linear variation close to the plates, and time averaged gradients in the bulk of the cell are small compared to the gradient in the vicinity of the plates. The existence of these regions of large gradient, the "boundary layers", can be interpreted as follows: since there is no source or sink of heat within the flow, the time averaged heat flux through every horizontal plane must be the same. Close to the plates, the advection velocity is small by virtue of the no slip condition at the solid surfaces, and the heat transport is mainly diffusive. Further away from the plates, however, mixing by turbulent velocity fluctuations leads to much more efficient heat transport than is possible by diffusion alone. The interior of the cell therefore acts like a thermal short circuit so that no large gradient can be sustained in the turbulent bulk. A simi-

Turbulence: A Tentative Dictionary
Edited by P. Tabeling and O. Cardoso, Plenum Press, New York, 1995

lar argument can be made for the velocity field, in which "momentum transport" stands in place of "heat transport". The velocity field is thus expected to have a boundary layer, too, in which the gradient of the mean velocity far exceeds the gradients found in the bulk of the cell.

The instabilities and subsequent detachments of the boundary layers give rise to plumes[3], which reach into the core of the cell while being advected by the main circulation[4, 5, 6].

Although a wealth of experimental facts has been accumulated, our understanding of convection is poor and models are controversial. Here we will present our experimental knowledge of convection at "high" Ra (roughly $Ra > 10^6$, where the flow is clearly turbulent) and its most basic interpretations. For a thorough review of the theory, we refer the reader to ref. [7]. The present paper is organized essentially according to the historical order in which various issues were addressed. We will first present a classification of turbulent convective flows and introduce some of the ideas that have been put forward to rationalize the behavior of the flow as a function of Ra. We will then review visual observations of the "coherent objects" (swirls and plumes) and more quantitative investigations of the boundary layers.

A global characterization of convection is the heat flux through the cell. This is usually given in terms of the Nusselt number Nu, defined as the ratio of the actual heat flux to the heat flux that would be observed in absence of fluid motion. Since the temperature drop across the cell is expected to occur in two boundary layers of thickness λ_T, in which the suitably normalized temperature varies linearly from 0 to $\Delta/2$ at the bottom plate and from $\Delta/2$ to Δ at the top plate, one has

$$Nu = \frac{L}{2\lambda_T}$$

This relation has been directly verified at $Pr = 0.7$[8].

Local characterization of the flow is possible by recording temperature time series at the center of the cell, for example. Standard data reductions consist in evaluating their power spectrum (Fourier transform squared of the time series) and histogram or probability distribution function (the probability to find a temperature between T and $T + dT$ at any time). A technique for measuring the average velocity v of the large scale flow places near a sidewall a pair of bolometers vertically separated by some small distance d[2]. The time recordings from these two bolometers are virtually identical except for a time delay τ. The ratio d/τ provides an estimate for v.

In a series of experiments performed with gaseous helium at low temperature[9, 10, 2, 11], it was found that Nu, v, and the rms of temperature fluctuations at the center vary as powers of Ra, for Ra greater than approximately 10^8. The exponents, but not the range of Ra over which the scalings are valid, are independent of the aspect ratio of the cell (aspect ratios 0.5, 1 and 6.7 were used). For instance, Nu scales as $Ra^{2/7}$ for $Ra > 10^8$ in all cells, but this particular law extends down to $Ra \approx 10^4$ in the aspect ratio 6.7 cell. v scales approximately as $Ra^{1/2}$, but the data could also be fitted to $Ra^{3/7}$ times logarithmic factors, as was proposed in ref. [12] . Furthermore, the time series measured in the center of the cell for $Ra > 10^8$ have power spectra which decay as $f^{-7/5}$ over one and a half decades of frequency f, and the histograms have a shape independent of Ra, but which depends on aspect ratio (exponential for aspect ratio 1, non-gaussian with exponential tails for aspect ratios 0.5 and 6.7). The state at high Ra has been called "hard turbulence", as opposed to "soft turbulence" for the regime at lower Ra, where the behavior is generally aspect ratio dependent and histograms measured in the center of the cell are gaussian.

Modifications in the shape of power spectra suggest the existence of another regime for $Ra > 10^{11}$[13]. However, it is presently a matter of debate whether these changes are inherent to the flow or just due to the response of the bolometers[14].

We will now interrupt the presentation of experimental results and describe some of the models related to the hard turbulence regime, in order to motivate the next round of experiments. Arguments have been given[15] that the spatial spectrum of the temperature field should decay like $k^{-7/5}$ as a function of wavenumber k for scales between the cell size and the dissipative scales, at which the dynamics is dominated by molecular diffusion. If one furthermore assumes that the "frozen flow hypothesis" is valid (meaning that the mean advection velocity is sufficiently large compared to the velocity fluctuations so that the temporal spectrum measured by a fixed probe is identical to the spatial spectrum), one predicts a range of frequencies over which the temperature power spectrum should vary as $f^{-7/5}$. However, the frozen flow hypothesis is extremely questionable in convection, since velocity fluctuations exceed the mean velocities in the central portions of the cell[16, 17], which makes the conversion of wavenumbers into frequencies problematic.

The Nusselt number measurements are at variance with earlier theories which predicted $Nu \propto Ra^{1/3}$[18]. The first argument in support of $Nu \propto Ra^{2/7}$ is presented in ref. [10] . The thermal boundary layer and the center of the cell are assumed to be separated by a mixing layer which allows for matching of the velocities and temperatures characteristic of the boundary and central regions. A series of dimensional arguments leads to the experimentally observed power law for Nu and the temperature fluctuations at the center of the cell in the hard turbulence regime.

It was later discovered that a large scale flow exists in the cell[2], which was unknown at the time the above model appeared. A more recent theory by Shraiman and Siggia[12] proposes that the large scale flow creates a turbulent boundary layer at the plates, whose dynamics tune the heat flux through the thermal boundary layer. By using empirical relations valid for turbulent boundary layers, the correct $Nu(Ra)$ and $v(Ra)$ are derived.

These two models share the view that scaling laws are ultimately determined by the interaction between the boundary layer and the center, and that they cannot be understood by considering any region of the cell in isolation. These theories motivate a closer investigation of the boundary layers and their interaction with the core. A recent approach[19, 14] attributes transitions of the convective regime to a crossing of relevant lengthscales in the experiment (the size of the cell, the size of the mixing layer, the distance of a detector from the plates), and the lengthscale below which isothermal surfaces wrinkle or perhaps are fractal. The Ra at which transitions ought to occur according to this theory are in good agreement with Rayleigh numbers, at which changes are experimentally observed in temperature power spectra. However, no predictions are made for the scalings of Nu and v.

We now return to the experiments. Some intuition about boundary layer dynamics can be gained from visualization. A particularly spectacular method suspends microencapsulated spheres of liquid crystal (typical size 50-100 μm) in water[4, 5, 6, 20]. If illuminated with white light, these liquid crystals appear in different colors depending on their temperature, so that a slice of the temperature field can be directly visualized. The picture of plume generation is as follows: in a cubic cell, the general circulation is oriented diagonally across the cell; vertically moving fluid hits the boundary layers in opposite corners of the cell. Wavelike perturbations start in these impact regions and propagate horizontally along the thermal boundary layer with the direction of the large scale flow. The crests of these perturbations may grow during propagation and turn into a swirl or a plume. In the corners where the circulation stagnates and is deflected from horizontal into vertical motion, buoyancy forces obviously dominate inertia and shear stress, which is not necessarily true at the center of the plate. Most plumes therefore detach from the boundary layer when they reach the edge of the plate, and move vertically along the sidewall instead of crossing the cell. Correspondingly, it has been measured in water that only a negligible fraction of the heat transport traverses the cell[17].

Water carried by the circulation, say, from the bottom to the top plate, does not thermalize during the ascent and arrives near the top plate with a mean temperature higher than the mean temperature at the center of the cell. The core of the cell is stably stratified! This stratification was observed both by visual methods[20] and local temperature measurements[17]. The gradient of the temperature inversion is around $2 \, 10^{-2} \Delta/L$ at $Ra = 10^9$ and $Pr = 6.6$. An inversion is however not found in gases, which shows that the heat exchange between the large scale flow and the center of the cell is more effective for lower Pr.

We lastly turn to a more quantitative description of the boundary layers. A complete characterization of the thermal and velocity boundary layers is possible in water, where velocities can be conveniently measured by tracking parcels of fluid labeled with electrochemically produced dye[17]. The mean velocity must be zero at the exact center of the cell by symmetry, and must also vanish at the plates due to the no slip condition, so that the mean velocity must reach a maximum somewhere in between. Two lengthscales are therefore of interest when dealing with the velocity profile: the thickness of the viscous boundary layer λ_v, within which the mean horizontal velocity varies linearly, and the distance L_v at which the velocity is maximum. At $Ra = 10^9$ and $Pr = 6.6$, it was found that $\lambda_v \approx 2\lambda_T$ and $L_v \approx 5\lambda_T$. The Reynolds number based on L_v is around 50, which indicates that the boundary layer is not turbulent in the usual sense. Not surprisingly, the veloc-

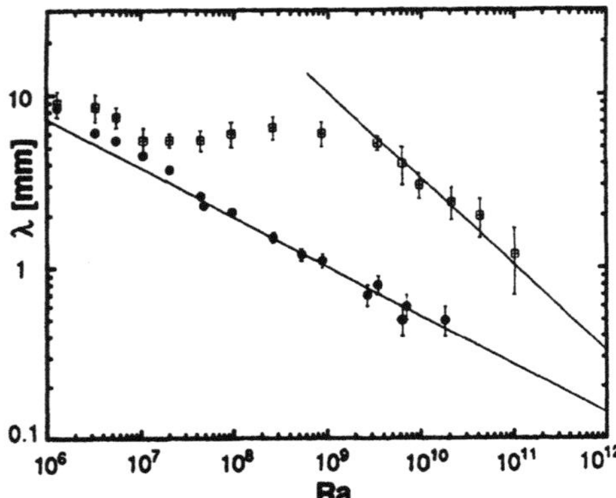

Figure 1: The thickness of the thermal boundary layer λ_T (circles) and the distance from the plate at which the cut off frequency of the temperature power spectrum is maximum (squares) as a function of Ra. The line drawn for λ_T corresponds to $Ra^{-0.29}$, the line drawn through the squares corresponds to $Ra^{-1/2}$.

ity profile is different than predicted by the "logarithmic law", which is well documented for shear layers at higher Reynolds numbers[21]. The major sources of instability must then be the unstable temperature stratification within the velocity boundary layer and instabilities of the large scale flow invading the boundary layers.

For temperature probes just outside the thermal boundary layer, the passage of plumes appears as spikes in the time series. From visualization, we know that the width of the stems of plumes does not vary appreciably with distance from the plates. Thus, the spikes in the temperature time series are narrowest (or conversely the spectrum extends to the highest frequencies) at the distance L_v from the plate. This correspondence between velocity and temperature power spectra has been verified in water and put to use in experiments with gases, for which direct velocity measurements are not practical.

The data in Fig. 1 have been obtained with a high pressure gas cell at room temperature[8]. This figure shows as a function of Ra the thickness of the thermal boundary layer λ_T and the location of the broadest temperature power spectrum.

Following the reasoning presented in the previous paragraph, this latter lengthscale approximately equals L_v. With this interpretation, three regimes clearly appear: for $Ra < 10^7$, λ_T and L_v change according to the same law. For $10^7 < Ra < 10^9$, L_v is approximately constant; it decreases again for $Ra > 10^9$. This decrease is compatible with a $Ra^{-1/2}$ law, although the range of Ra is too small to determine an exponent with confidence. A $Ra^{-1/2}$ scaling of L_v would mean that the Reynolds number of the velocity boundary layer (vL_v/ν) is constant. The average advection velocity can be estimated from delay measurements with a pair of thermistors (as described above for the helium experiments). One obtains a limiting value for the Reynolds number of 600 (as opposed to 50 measured in water), which is a reasonable value for a turbulent boundary layer. Note that Reynolds numbers are higher at lower Pr for equal Ra.

For $Ra > 2.10^7$, one finds $\lambda_T \propto Ra^{-2/7}$, as expected from Nu measurements for the "hard turbulence" regime. Extrapolation of the data in Fig. 1 predicts a crossing of λ_T and L_v at approx $Ra \approx 10^{14}$, above which a new regime of convection may be awaiting discovery. An interpretation of the range $10^7 < Ra < 10^9$ has been elusive so far.

These new experimental results will hopefully stimulate further theoretical work to model more faithfully the dynamics of the boundary layers and their interaction with the core of the cell. Future experimental investigation will continue in the direction of higher Ra, but also to smaller and faster detectors, and to low Pr (liquid metals), about which so little is known. Finally, new techniques are being developed for the visualization of spatially extended velocity or temperature fields[20, 22].

Bibliography

[1] R. Krishnamurti, and L.N. Howard, Proc. Natl. Acad. Sci. USA **78**, 1981 (1981)

[2] M. Sano, X.Z. Wu, and A. Libchaber, Phys. Rev. A **40**, 6421 (1989)

[3] E.M. Sparrow, R.B. Husar, and R.J. Goldstein, J. Fluid Mech. **41**, 793 (1970), J.S. Turner, "Buoyancy Effects in Fluids", Cambridge University Press 1973

[4] T.H. Solomon and J.P. Gollub, Phys. Rev. Lett. **64**, 2382 (1990)

[5] T.H. Solomon and J.P. Gollub, Phys. Rev. A **43** 6683 (1991)

[6] G. Zocchi, E. Moses, and A. Libchaber, Physica A **166**, 387 (1990)

[7] E. Siggia, to appear in Ann. Rev. Fluid Mech. 1994

[8] A. Belmonte, A. Tilgner, and A. Libchaber, Phys. Rev. Lett., **70**, 4067 (1993)

[9] F. Heslot, B. Castaing, and A. Libchaber, Phys. Rev. A **36**, 5870 (1987)

[10] B. Castaing, G. Gunaratne, F. Heslot, L. Kadanoff, A. Libchaber, S. Thomae, X.Z. Wu, S. Zaleski, and G. Zanetti, J. Fluid Mech. **204**, 1 (1989)

[11] X.Z. Wu and A. Libchaber, Phys. Rev. A **45**, 842 (1992)

[12] B.I. Shraiman, and E.D. Siggia, Phys. Rev. A **42**, 3650 (1990)

[13] X.Z. Wu, L. Kadanoff, A. Libchaber, and M. Sano, Phys. Rev. Lett. **64**, 2140 (1990)

[14] S. Grossmann and D. Lohse, Phys. Lett. A **173**, 58 (1993)

[15] I. Procaccia and R. Zeitak, Phys. Rev. Lett. **62**, 2128 (1989), V.S. L'vov, Phys. Rev. Lett. **67**, 687 (1991), V. Yakhot, Phys. Rev. Lett. **69**, 769 (1992)

[16] A.M. Garon and R.J. Goldstein, Phys. Fluids **16**, 1818 (1973)

[17] A. Tilgner, A. Belmonte, and A. Libchaber, Phys. Rev. E **47**, R2253 (1993)

[18] W.V.R. Malkus, Proc. R. Soc. Lond. **A225**, 185 (1954); L.N. Howard, J. Fluid Mech. **17**, 405 (1963)

[19] I. Procaccia, E.S.C. Ching, P. Constantin, L.P. Kadanoff, A. Libchaber, and X.Z. Wu, Phys. Rev. A **44**, 8091 (1991)

[20] B.J. Gluckman, H. Willaime, and J.P. Gollub, Phys. Fluids A **5**, 647 (1993)

[21] Tennekes and Lumley, "A First Course in Turbulence", MIT Press 1972

[22] R.J. Adrian, R.D. Keane, and P.W. Offut, in Proceedings of the "Workshop on Visualization and Statistical Analysis in Hard Turbulence" , Sept. 9-10, 1992, Minneapolis, Minnesota, p. 3

SCALING IN HYDRODYNAMICS

L. P. KADANOFF

The Research Institutes, The University of Chicago,
5640 S. Ellis Avenue, Chicago Illinois, 60615, USA

The concept of scaling and dimensional analysis has been extensively used in hydrodynamics, and most especially in the study of turbulence. But, in fact, scaling is used in many different ways in this subject.

DIMENSIONAL ANALYSIS

At its simplest, scaling describes the elimination of dimensional quantities and the isolation of dimensionless combinations which then describe the situation in question. Let me describe the simplest example: the behavior of the Navier Stokes equation

$$u_t + (u \cdot \nabla)u = -\frac{\nabla p}{\rho} + \nu \nabla^2 u$$
$$\nabla \cdot u = 0 \qquad (1)$$

in a situation in which one is looking at a flow characterized by a typical magnitude of the velocity, U, and a typical length scale L. If one then uses the dimensionless quantities for velocity $v = u/U$, for length $R = r/L$, and for time $T = tU/L$ then the Navier-Stokes equation becomes

$$v_T + (v \cdot \nabla)v = -\nabla P + Re\nabla^2 v$$
$$\nabla \cdot v = 0 \qquad (2)$$

In Eq. (2) the gradient is a derivative with respect to the rescaled space variable, R, while P is a rescaled pressure defined to keep the velocity while the Reynolds number, Re, is defined by

$$Re = \frac{UL}{\nu} \qquad (3)$$

Since the kinematic viscosity is quite small for most fluids (for example being $1.0 \times 10^{-6} m^2/sec$ for water and $1.5 \times 10^{-5} m^2/sec$ for air) the Reynolds number is often quite high.

One uses this kind of scaling by saying that for a fixed geometry the dimensionless flow only depends upon the Reynolds number. Thus, for example, in the picture book by Van Dyke[1], many examples are given of flow past a sphere. Neither the velocity at infinity, U, nor the radius of the sphere, L, are specified but only the Reynolds number. For low values of Re the flow is laminar, and then as Re is increased the flow gets more complex and unsteady, until at still larger values it can be said to be truly turbulent.

In the study of convective flow, the Boussinesq equations are also converted into dimensionless form. The physical temperature variable, T, is rescaled by a typical temperature difference in the system, Δ, to get the dimensionless combination $\theta = T/\Delta$. Then the relevant quantities obey rescaled equations in which there is two dimensionless parameters, the Rayleigh number:

$$Ra = \frac{g\alpha\Delta L^3}{\kappa\nu} \qquad (4)$$

and the Prandtl number

$$Pr = \frac{\nu}{\kappa} \qquad (5)$$

Here κ is the thermal diffusivity, g the acceleration of gravity, and α the thermal expansion coefficient of the fluid. The Rayleigh number measures how hard the fluid is being 'pushed' so that turbulence occurs for large values of Ra.

SIMILARITY SOLUTIONS

This 'naive' dimensional analysis is only the simplest example of the kind of scaling analysis used in the field. Often the scaling is based upon a relatively sophisticated view of similarity solutions. The simplest and most classical analysis of this kind is due to Blasius and treats the flow past a flat plate.

To see the structure of the result take the curl of the Navier-Stokes equation to find

$$(u \cdot \nabla)\nabla \times u = \nu \nabla^2 \nabla \times u \qquad (6)$$

The Blasius solution is based upon the idea that there is a characteristic scale of the y coordinate, which varies with x, the distance along the plate measured from its beginning. Therefore, every function of y will depend upon y in the form of functions of $y/Y(x)$ where $Y(x)$ is the characteristic scale of y. The scale of x is simply the distance from the leading edge of the plate. We look far downstream, where the scale of Y is much smaller than the scale of x. In this limit, one can neglect the $(\partial/\partial x)^2$ term in ∇^2 in comparison with the $(\partial/\partial y)^2$ term. Thence (6) reduces to

$$(u \cdot \nabla)\nabla \times u = \nu \frac{\partial^2}{\partial y^2}\nabla \times u \qquad (7)$$

with the additional condition that the divergence of the velocity vanishes.

To an order of magnitude one can estimate the size of a y-derivative as being proportional to $Y(x)^{-1}$, while an x-derivative is of the order of the inverse x-scale x^{-1}. A typical value of u_x is just U. Thus, if the two terms in Eq (7) are to balance out, we must have that, to an order of magnitude:

$$\frac{U}{x} \sim \frac{\nu}{Y(x)^2}$$

so that the y-scale is

$$Y(x) = \left(\frac{\nu x}{U}\right)^{1/2} \qquad (8)$$

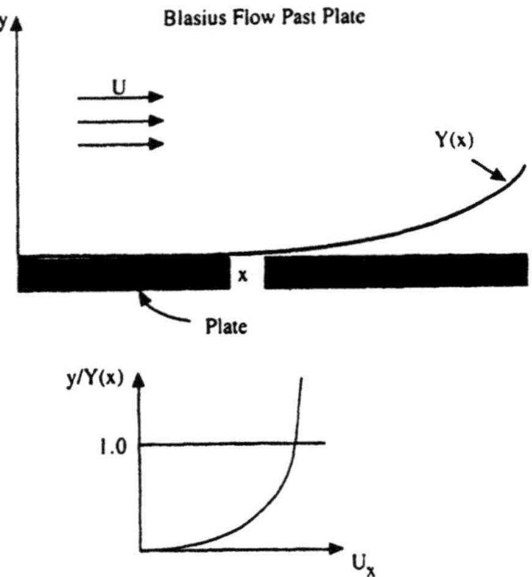

Figure 1 : Blasius Figure. This figure shows first the geometry of the situation described by Blasius and then the form of the velocity profile achieved.

Using this idea, one can in fact, construct a solution of (7) assuming that the x-component of the velocity has the form

$$u_x = U\Psi\left(\frac{y}{Y(x)}\right) \qquad (9)$$

with Ψ obeying suitable boundary conditions.

The physical idea upon which this is all based is that the flow is the same for all x except for the change of scale, represented by the $Y(x)$. Notice that the answer can be represented by simple power laws, for example that $Y(x)$ is proportional to $x^{1/2}$. We see here that the simple power laws are an outcome of the idea that as one goes to larger x nothing changes except the scale of y.

INERTIAL RANGE SCALING

There is a third use of scaling in turbulence theory based upon the notions of Kolmogorov. In 1941[2], he imagined that turbulence could be described as the flow of energy from large scales to small. He visualized the energy to be injected at a rate ϵ as a large kind of eddy. Then the kinetic energy would respond to the non-linear term $(u \cdot \nabla)u$ in the hydrodynamic equation by forming successively smaller eddies and thereby carry kinetic energy down to smaller and smaller scales. Finally, as the eddies became really small the viscous term in the hydrodynamic equation would become important and the kinetic energy would be converted into heat.

This argument could then be converted into a kind of dimensional analysis. One imagines that the velocity difference over a distance of order r is a dimensional quantity, u_r, depending on the distance scale, times a stochastic variable, ξ. In symbols,

$$u(R + r) - u(R) = u_r \xi(R, r) \qquad (10)$$

Here, ξ, is of order unity. Dimensional analysis gives the current of kinetic energy from the scale r to the next scale as the kinetic energy density at that scale, u_r^2, times the flow rate, u_r/r. The net result is that the downward flow of kinetic energy is given by

$$\epsilon_r = \frac{u_r^3}{r} \qquad (11)$$

Since the kinetic energy is viewed to flow from scale to scale until it disappears, that quantity should be independent of r until the scale of r becomes small enough so that viscosity and dissipation become important. If ϵ is the common value of all the ϵ_r then Eq (11) gives the scale dependence of the velocity as:

$$u_r = (\epsilon r)^{1/3} \qquad (12)$$

Eq (12) represents a kind of *scaling law* which describes how the magnitude of a fluctuating quantity depends upon the distance scale. Later on, considerably after Kolmogorov's work, this concept became fashionable in critical phenomena[3].

The behavior (12) should describe the entire inertial range of behavior:

$$\eta << r << L \qquad (13)$$

Here, L is the size of the scale on which energy is injected. This is called the inertial scale. The velocity on this scale, U, is related to ϵ by the direct extension of Eq (12)

$$U = (\epsilon L)^{1/3}$$

On the other hand, η is the scale on which energy is removed by dissipation. At this scale, the viscous term and the non-linear term in the Navier-Stokes equation are of the same order so that

$$\frac{u_r^2}{r} \sim \nu \frac{u_r}{r^2}$$

We should not be surprised to learn that this equation implies that the Reynolds number at this scale should be of order unity, as a consequence, the dissipative scale is determined by the condition that the dissipative scale is given in terms of the Reynolds number of Eq (3) by

$$\eta = L Re^{-3/4}$$

To see the implications of these ideas, imagine that one measures the velocity of fluid in a sensor at rest at a particular position within a turbulent fluid. Nonetheless fluid is being carried past this sensor at a typical velocity U, since the entire fluid is being carried with this large scale velocity.

$$V(\omega) = \int_0^T dt\, u(t) e^{i\omega t} \qquad (14)$$

and form $P(\omega)$ to be proportional to the squared magnitude of Fourier transform

$$P(\omega) \sim \frac{< |V(\omega)|^2 >}{T} \qquad (15)$$

This $P(\omega)$ is called the power spectrum. Here $<>$ denotes an ensemble or spatial average. The averaged result is independent of T.

At this point, we would like to develop the consequences of the Kolmogorov theory for the power spectrum. This involves using a form of what is called the Taylor[†] frozen flow hypothesis. Since everything is being carried past the sensor with a typical speed U, according to the Kolmogorov theory, a measurement at the frequency ω should entail spatial scales comparable to U/ω. Thus the two factors of

---

[†] After G. I. Taylor

u_r in expression (14) and (15) lend expression (15) a factor of $U^2(r/L)^{2/3}$, which is $\omega^{-2/3}L^{-2/3}U^{8/3}$. The two time integrals in Eq (15) respectively lend a factor of T, to cancel out the explicitly T in Eq (15) and also a factor of ω^{-1}, to make everything dimensionally correct. The net result is that Eq (15) becomes

$$P(\omega) \sim \omega^{-5/3}L^{-2/3}U^{8/3} \qquad (16)$$
$$\text{for} \qquad \frac{U}{L} << \omega << \omega_D$$

which holds in the inertial range of frequencies corresponding to the limits (13). The result is that the dissipative scale, ω_D, corresponds to a highest frequency for inertial range behavior, determined as

$$\omega_D \sim \frac{U}{\eta} \sim \left(\frac{U}{L}\right) Re^{3/4} \qquad (17)$$

A generalization of this result can be used to extend the form into the dissipative region, by saying that in this region the power spectrum depends only on the dimensionless variable which gives the ratio of frequency to dissipative frequency, i.e.

$$P(\omega) \sim \omega^{-5/3}L^{-2/3}U^{8/3}p^* \left(\frac{\omega}{\omega_D}\right) \qquad (18)$$
$$\text{for} \qquad \frac{U}{L} << \omega$$

In Monin and Yaglom[4] this theoretical result is compared with experiment. The fit is quite good indeed.

Bibliography

[1] M. Van Dyke, *An Album of Fluid Motion*, The Parabolic Press, P.O. Box 3032, Stanford California, 94305-0030 USA (1982).

[2] A.N. Kolmogorov, Doklady Acad. Sci. U.S.S.R. **31** 6 (1941), **32** No 1 (1941).

[3] See for example B. Widom, J. Chem. Phys. **43**, 3892,9898 (1965) and L.P. Kadanoff, Physics **2**, 263 (1966).

[4] See A.S. Monin and A.M. Yaglom, Statistical Fluid Mechanics, The MIT Press, Volume 2, figure 75.

Shear Flows
(Turbulent)

Charles W. Van Atta
Department of Applied Mechanics and Engineering Sciences
and Scripps Institution of Oceanography
University of California, San Diego
La Jolla, California, 92093, U.S.A

Turbulent Shear Flows are turbulent flows in which the mean velocity varies as a function of one or more of the three spatial coordinates so that there are gradients (shear) in the mean velocity profiles.

Turbulent shear flows are ubiquitous in nature and in applications in physics and technology. Some examples in nature are the flows associated with the Great Red Spot of Jupiter and the shorter lived hurricanes on Earth, wind profiles in the atmospheric boundary layer, the atmospheric jet stream, and the Equatorial Undercurrents in the Atlantic and Pacific oceans. Engineering examples include rocket exhaust jet flows, flow in oil pipelines, boundary layer flows on aircraft and submarines, and fluid dynamical lasers. From the point of view of energetics of the Reynolds averaged energy equation the basic difference between unsheared turbulent flows like grid generated turbulence and shear flows is that the presence of mean shear coupled with the Reynolds stress gives a source of turbulent kinetic energy, which is formally equal to the product of the Reynolds stress and the gradient of the mean velocity.

Here we will consider some of the salient features of several different turbulent shear flows to illustrate some of their physical aspects. There are many other interesting and important turbulent shear flows besides the ones mentioned here, including flows influenced by rotation, magnetic fields, curvature effects, and so on. Each different kind of flow invariably introduces new effects and turbulence structures peculiar to itself, prompting the common observation that there is no single *"turbulence problem" per se*, there are only turbulent *flows* and each one is a separate problem.

HOMOGENEOUS TURBULENT SHEAR FLOW

In homogeneous turbulent shear flows, the mean velocity U in the x direction is a linear function of the lateral coordinate z, i.e. $U(z) = az$. All statistical quantities are independent of z. Experiments show that for a nonstratified, homogeneous fluid the turbulence production exceeds the dissipation and the velocity fluctuations grow spatially with x without bound, roughly exponentially with x. Similar behavior is found in direct numerical simulations in which the turbulence develops in time rather than spatially. If the flow is stably stratified, Rohr *et al.* (1988) found that the growth rate of the turbulence is a decreasing function of the Richardson number for $R < R_{cr}$, and becomes negative for $R > R_{cr}$, with R_{cr} equal to roughly 0.3. For $R = R_{cr}$ the turbulence production, dissipation, and buoyancy sink term which produces potential energy are in balance, and the energy in the velocity and density fluctuations is constant. There is, however, a continual spectral redistribution from small to large scales taking place. Direct Numerical Simulations (DNS) of Holt *et al.* (1992) suggest that the critical Richardson number might be a function of Reynolds number, but this has not so far been seen experimentally. DNS results also suggest that there are coherent structures in homogeneous shear flows in the form of horseshoe-like vortices oriented nearly along the principal extensional strain direction and that the effect of stable stratification in reducing the vertical transport is associated with a weakening of these vortical structures.

WALL BOUNDED TURBULENT SHEAR FLOWS

In Turbulent Boundary Layers and Pipe or Channel Flows, turbulent momentum and heat transfer produce large mean velocity and temperature gradients at the surface and hence large drag and heat transfer compared with laminar flows. The strong correlations between velocity components necessary for this efficient momentum, heat, and mass transport are produced by coherent structure "bursting" cycles or other vortical events of various types. In one scenario spanwise vorticity associated with sublayer "streaks" near the wall is perturbed into vortex loops or horseshoes whose legs then contain streamwise and wall-normal vorticity (Theodorsen, 1962). These vortex structures then move by self-induction out into the boundary layer where they are further stretched until they "burst" and disappear. The largest contributions to the Reynolds stress occur during the "ejection" phases of the burst cycle, during which the streamwise and wall-normal velocity fluctuation components are strongly anti-correlated. This and other bursting scenarios are discussed in a recent review of coherent motions in the boundary layer by Robinson (1991). In terms of Reynolds averaged statistics, over much of the inner region of the boundary layer, the turbulence production is nearly balanced by the dissipation, and in this region the mean velocity is a logarithmic function of distance from the bounding surface.

FREE TURBULENT SHEAR FLOWS

The Turbulent Free Shear Layer between two streams of differing velocities spreads linearly downstream and a corresponding similarity exists in which dimensionless mean quantities depend only on the ray angle from an apparent origin. For quiet initial conditions the coherent vortices associated with the initial Kelvin-Helmholtz instability persist to become the initial coherent structures in the fully developed turbulence. These nearly two-dimensional structures amalgamate by pairing (and sometimes tripleting) downstream so that their size grows linearly along with the layer thickness. At a certain distance downstream smaller vortices with their axes wrapped around the K-H rollers appear. These vortices have vorticity components in all three directions, but are commonly called "longitudinal" or "streamwise" vortices because of their appearance in a planview of the flow. These vortices catalyze a "mixing transition" beyond which the lateral mixing is considerably enhanced. If the density of the upper stream is less than that of the lower (stably stratified) new types of *"convective"* instabilities also come into play, as described by Schowalter *et al.* (1994a,b). As the interface between the two fluids is rolled up by the K-H instability, along some parts of the interface one finds light fluid above heavy (stable) and here the formation of "streamwise" vortices is inhibited. At other sections of the interface where there is heavy fluid above light (unstable) convective instability driven by baroclinic generation of vorticity also produces vortex pairs wound around the K-H rollers and these vortices then dominate the transition process.

A conceptual model of Broadwell and Breidenthal (1982) examines the rate at which a turbulent shear layer actually molecularly mixes the different species in the two streams, which depends on the rates of entrainment, macroscopic deformation, and molecular diffusion. Entrainment is the slowest step unless $D << \nu$, where D is the molecular diffusivity and ν is the kinematic viscosity. In gases, $D/\nu \sim 1$ and entrainment is the bottleneck. In liquids, however, ν may exceed D by a factor of hundreds or more, so that entrainment and diffusion may both provide important constraints on the mixing rate.

Resolving scalar turbulence down to molecular diffusion scales is difficult in liquids with $D << \nu$. However, using optical techniques the fine scale structure of a high Schmidt number (Schmidt number $S = \nu/D$) scalar field mixed in a turbulent jet has been successfully examined by Dahm *et al.* (1991) using a dynamically passive dilute laser fluorescent dye.

The formation and transition regions of jets and wakes exhibit coherent behavior, even at very large Reynolds numbers. The presence or absence of vortex like coherent structures in the self-similar region, and their topology, is not yet clear. In plane jets and wakes the instability that, in shear layers, leads to vortex pairing is greatly reduced in the staggered parallel rows of opposite sign vortices on opposite sides of the flow. In round jets and wakes the absence of obvious large scale coherent motions may be associated with the inherent, violent instability of pairing vortex rings. However, such flows are unstable to disturbances that gather vorticity up into a helix, rather than a succession of separate rings (see Perry and Lim 1978).

THE QUESTION OF LOCAL ISOTROPY IN TURBULENT SHEAR FLOWS

According to Kolmogorov's original ideas, one should expect to see local isotropy in the small scales in a turbulent shear flow, provided that the scale separation between large and small scales is sufficiently large. "Local isotropy" means isotropic behavior over a certain restricted or "local" wavenumber range. The relations for locally isotropic behavior of spectra, gradient moments, etc., are derived purely from kinematical and continuity arguments, not from dynamics. A large

Reynolds number is not necessary for local isotropy, but the extent of the region of local isotropy is expected to increase with Reynolds number. The smallest scales in a low Reynolds number turbulent flow often exhibit local isotropy. However, it seems reasonable that if the ratio of mean strain to fluctuating strain is large enough one would expect anisotropic behavior. For energy spectra of the velocity and scalar fluctuations, Kolmogorov's idea appears to be borne out. Experiments and DNS at low R_λ show isotropic behavior at the smallest scales only, while experiments in the laboratory, atmosphere, and ocean show that the extent of isotropy increases as R_λ increases. For a review and recent laboratory experiments at high R_λ see Praskovsky *et al.* (1993). For moments of velocity and scalar gradients the available data show a marked degree of anisotropic behavior. This is at first puzzling, as taking gradients is expected to emphasize the contributions of the fine structure. However, as is well known for nondifferentiated velocity and temperature fluctuations, such moments do include contributions from all scales, can be very sensitive to sharp gradients found near the edges of large scale coherent structures, and are generally not suitable for examining particular wave number ranges for *locally* isotropic behavior. The degree of local isotropy may be unambiguously examined by comparing the relative behavior of *spectra* of the gradients directly with the appropriate relations for local isotropy, as discussed in Van Atta (1991). This has yet to be done for turbulent shear flows, as most studies to date have measured only moments and not the spectra of gradients. An exception is the work of Dahm *et al.* (1991) mentioned earlier, in which it was found that the spectra of the small scale scalar gradient field (all three components were measured) obeyed isotropic relations within the uncertainty of the somewhat limited statistical sample.

Bibliography

[1] Broadwell, J.E. and Breidenthal, R.E. 1982 A Simple Model of Mixing and

Chemical Reaction in a Turbulent Shear Layer, J. Fluid. Mech. **125**, 397-410.

[2] Dahm, W.J.A., Southerland K.B., and Buch, K.A. 1991 Direct, high resolution, four-dimensional measurements of the fine scale structure of $Sc \gg 1$ molecular mixing in turbulent flows, Phys. Fluids A **3**(5), 1115-1127.

[3] Holt, S.E., Koseff, J.R., and Ferziger, J.H. 1992 A numerical study of the evolution and structure of homogeneous stably stratified sheared turbulence, J. Fluid. Mech. **237**, 499-539.

[4] Perry, A.E. and Lim, T.T. 1978 Coherent structures in coflowing jets and wakes, J. Fluid. Mech. **88**, 451-464.

[5] Praskovsky, A.A., Karyakin, Yu M., and Kuznetsov, V.R. 1991 Experimental verification of local isotropy assumption in high Reynolds number flows. In Proc. Third European Turbulence Conference, Stockholm 1990.

[6] Robinson, S.K. 1991 Coherent Motions in the Turbulent Layer, Ann. Rev. Fluid Mech. **23**, 601-39. Rohr, J.J., Itsweire, E.C., Helland, K.N., and Van Atta, C.W. 1988 Growth and decay of turbulence in a stably stratified shear flow. J. Fluid Mech. **195**, 77-111.

[7] Schowalter, D.G., Van Atta, C.W. and Lasheras, J.C. 1994a A study of streamwise vortex structure in a stratified shear layer, submitted to J. Fluid Mech.

[8] Schowalter, D.G., Van Atta, C.W., and Lasheras, J.C. 1994b Baroclinic generation of streamwise vorticity in a stratified shear layer, Meccanica (in press).

[9] Theodorsen, T. 1962 Mechanism of Turbulence, Proc. 2nd Midwestern Conf. on Fluid Mech., Columbus, Ohio, 1-18.

[10] Van Atta, C.W. 1991 Local isotropy of the smallest scales of turbulent scalar and velocity fields. Proc. Roy. Soc. London Ser. A, **434**, 139-147.

Shell Model
(of Turbulence)

M.H. Jensen
The Niels Bohr Institute
Blegdamsvej 17, DK-2100 Copenhagen Ø, Denmark
G. Paladin
Dipartimento di Fisica, Università dell'Aquila
Via Vetoio, I-67010 Coppito L'Aquila, Italy
A. Vulpiani
Dipartimento di Fisica, Università "La Sapienza"
P.le Aldo Moro 2, I-00185 Roma, Italy

The idea behind shell models of turbulence is to formulate a reasonable model that mimics the Navier-Stokes equations

$$\partial_t u + (u \cdot \nabla)u = -\frac{\nabla P}{\rho} + \nu \Delta u + f_u \quad (1)$$

and with which it is possible to study numerically, and even analytically, the fundamental properties of turbulent flows, like the energy cascade, probability distributions, intermittency effects, corrections to Kolmogorov theory, to mention a few. In practice, a direct numerical simulation of Eqs.(1) can only be performed at moderate Reynolds numbers (at the limit of the present computer facilities), and, in any case, it is highly non-trivial to extract such detailed information from it. For shell models on the other hand, it is possible to do numerical studies at very high Reynolds numbers ($\sim 10^{10}$) and to obtain an extended inertial range for the energy cascade. The basic idea of a shell model is that, instead of considering the Fourier transform of Eqs.(1) in a continuous k-space, one formulates it on a *discrete* set of wavevectors in k-space, thereby constructing an ordinary differential equation on each shell. The central part of the approximation is to write down the form of the coupling term between the various shells, i.e. to assume local or global interactions and decide how many terms from other shells should be included in the ODE for a particular shell. In general shell models contain the basic features of the physical properties of the fluids. Standard shell models have a relatively small number of degrees of freedom $O(10^2)$ so that they can be analyzed by the well known techniques of dynamical systems[1, 2]. The set of ODE is derived under the assumption that the most relevant mechanism for the behavior of the velocity field u is given by cascade transfers from large to small scales. To be more precise, the main hypothesis is that in the Fourier spectrum for u there is a (forward) cascade of the quantity u^2 towards the large wavenumbers, as a consequence of the advection term in Eqs.(1), which induces a transfer among different Fourier components of the velocity field.

Various shell models have been introduced for different turbulent phenomena[1-9]. It turns out that these models reproduce the main features of the small scale statistics of the velocity field at very high Reynolds number Re. Note that the number of degrees of freedom of the original Eq.(1) roughly increases as $Re^{9/4}$.

We discuss here a shell model which has been introduced by Gledzer[4], and developed by Ohkitani and Yamada[7, 8]. Models for the passive scalar advection[2], magnetohydrodynamics and plasmas[6] have been constructed using similar ideas. First, one divides the Fourier space in N shells. Each shell k_n ($n = 1, 2, .., N$) consists of the wavenumbers with modulus k such that $k_0 2^n < k < k_0 2^{n+1}$. The velocity increments $|u(\ell) - u(x + \ell)|$ on scale $\ell \sim k_n^{-1}$ are given by the complex variables u_n. The evolution equations are obtained according to the following criteria:

(a) the linear term for u_n is given by $-\nu k_n^2 u_n$

(b) the non-linear terms for u_n are a combination of the form $k_n u_{n'} u_{n''}$

(c) n' and n'' are nearest and next nearest neighbors of n

(d) in absence of forcing and damping one has conservation of volume in phase space (Liouville theorem) and the conservation of energy $\frac{1}{2} \sum_n |u_n|^2$

(e) the ('non-intermittent') scaling law of Kolmogorov $E(k) \sim k^{-5/3}$, i.e. $u_n \sim k_n^{-1/3}$, is a fixed point of the inviscid unforced evolution equations.

The properties (a), (b) and (d) are also valid for (1) in Fourier space, while (c) is an assumption on the locality of interactions among modes, which is rather well founded as long as $1 < \alpha < 3$ in the power law for the energy spectrum $E(k) \sim k^{-\alpha}$. (e) is a central ingredient of the model: it seems necessary to obtain scaling laws close to those derived by dimensional arguments. The instability of the fixed point is also very important. In fact, intermittency and corrections to the classical exponents of the scaling arise as a consequence of the motion of the system on a chaotic attractor in the $2N$ phase space.

The equations of our cascade model with N shells can be written as[1, 2]

$$(\tfrac{d}{dt} + \nu k_n^2) u_n = \; i k_n (\; a_n u_{n+1}^* u_{n+2}^* \\ + \tfrac{b_n}{2} u_{n-1}^* u_{n+1}^* \\ + \tfrac{c_n}{4} u_{n-1}^* u_{n-2}^*) \\ + f \delta_{n,4} \qquad (2)$$

with boundary condition

$$b_1 = b_N = c_1 = c_2 = a_{N-1} = a_N = 0$$

and where u_n are complex variables, ν is the viscosity, and f is an external forcing (here on the fourth mode). The conservation of phase space (for $\nu = f = 0$) is automatically satisfied by the absence of diagonal terms proportional to u_n in the right hand side of Eqs.(2). The coefficients of the non-linear terms should obey the relation $a_n + b_n + c_n = 0$ to satisfy the conservation of $\sum_n |u_n|^2$ (energy) in the absence of forcing and with $\nu = 0$. Moreover, they are defined a part a multiplicative factor (related to a time rescaling), so that one can fix $a_n = 1$. It follows that there is a free parameter Φ and one has

$$a_n = 1, b_n = -\Phi, c_n = -(1 - \Phi) \qquad (3)$$

where $0 < \Phi < 1$ if one requires that energy is the only conserved quantity. A direct calculation shows that for $\Phi > 1$ the symmetries of the model change since the inviscid unforced Eq. (2), beside energy, conserve the quantity $\sum k_n^\gamma |u_n|^2$ with $\gamma = -\ln_2(\Phi - 1)$. For instance, the direct cascade of enstrophy in 2D turbulence can be studied by the choice $\Phi = -5/4$. It is easy to check that Eqs.(2) with $\nu = f = 0$ have a fixed point which follows the Kolmogorov scaling $u_n \sim k_n^{-1/3}$, as well as a fixed point corresponding to the scaling $u_n \sim k_n^{.5 \ln_2(\Phi - 1)}$.

It is remarkable that for $0 < \Phi \leq 0.38..$, the Kolmogorov fixed point is stable in presence of viscosity and forcing. For larger Φ and in particular for the standard value $\Phi = 1/2$, the time evolution of the dissipative system (2) is chaotic and confined on a strange attractor in the $2N$ dimensional phase space. This fact is a strong evidence that the interaction between shells plays a fundamental role in determining the strength of the intermittency, and that the correct symmetries still leave a large freedom to the system (it can transfer energy either in a multifractal, intermittent way or in a Kolmogorov way).

In the following we shall describe the shell model of Yamada and Ohkitani with $\Phi = 1/2$ in (3) which exhibits an intermittent energy dissipation similar to that ob-

102

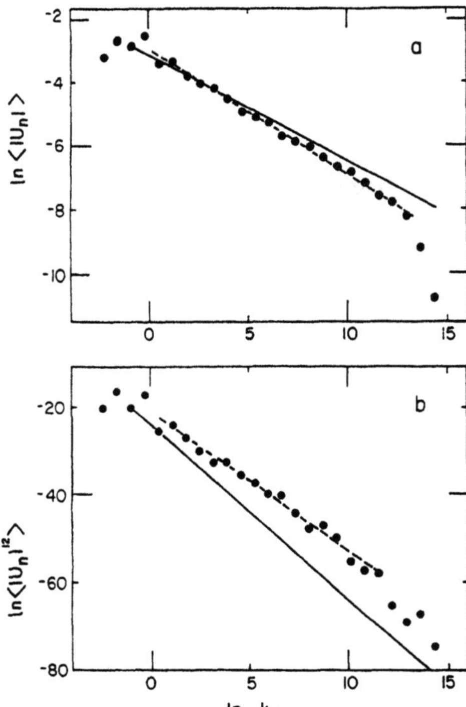

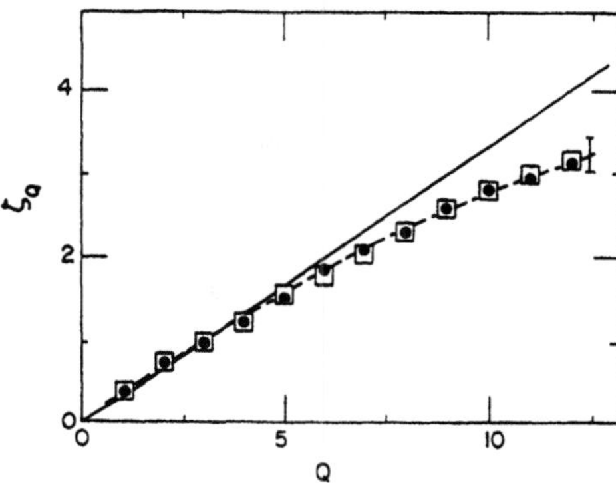

Figure 2 : The structure functions exponents $\zeta(Q)$ plotted versus Q. The data are obtained by an integration of Eqs.(2) with $N = 27$ and $N = 19$. The dashed line is the random β model fit[10] for the ζ_Q.

Figure 1 : Plot of the structure functions, $\ln< |u_n|^Q >$ versus $\ln k_n$ for a): $Q=1$ and b): $Q=12$ found from simulations of Eqs.(2) with $N=27$ over 10^4 time units. The full lines have slopes respectively -1/3 and -4 from Kolmogorov scaling. The dashes lines have slopes -0.39 and -3.18, respectively[1].

served in 3D fluids. To calculate the structure functions, let us consider the moments of the velocity differences over small length scales

$$\left\langle |\, \delta u(\ell)\, |^Q \right\rangle \sim l^{\zeta(Q)}. \qquad (4)$$

Instead of estimating a spatial average, the velocity differences are in the shell model averaged over time and Ref. [1] obtained structure functions up to the order $Q = 12$, by a numerical integration of Eqs.(2) with $N=19$ and 27 shells. Fig. 1 shows a plot of $\ln < |u_n|^Q >$ versus $\ln k_n$ for the two cases, $Q=1$ and $Q=12$. There is a good scaling over about five decades between the shell of the energy input and the shell of the Kolmogorov inner scale which defines the inertial range. The slope is $-\zeta(Q)$

and is *different* from the Kolmogorov result $< |u_n|^Q >\propto k_n^{-Q/3}$ which is indicated by the full drawn line. Fig. 2 collects the results for the structure function exponents with $N=27$, $\nu = 10^{-9}$ and $N = 19$, $\nu = 10^{-6}$. We note that the exponent $\zeta(Q)$ is not linear in Q[1]. The results are in very good agreement with the experimental results by Anselmet *et al* obtained from velocity measurements in a channel flow[11]. These data suggest an underlying multifractal structure of the energy dissipation since $\zeta(Q)$ can be obtained by the random β model[10], as shown by the dotted line in Fig.2. It is worth stressing that there exists a closure theory for the shell model which is able to reproduce *analytically* the values of the exponents ζ_Q[12].

One also observes that the intermittent behavior of the shell model is very close to the experimental results, even at a quantitative level. For example, the probability distribution function (PDF) of the velocity gradients exhibits the typical stretched exponential tails which are in good agreement with the prediction of the random β model[13] and direct numerical simulations[14].

The shell model also appears useful for

the analysis of the intermittent behavior by means of standard methods of the theory of dynamical systems. It allows one to relate the basic mechanisms of the energy cascade and the fluctuations of energy dissipation to the dynamical evolution. For instance, it is possible to compute numerically the spectrum of Lyapunov exponents and corresponding Lyapunov eigenvectors. It was found that during an intermittent burst there is a big increase of the chaoticity of the model and a localization of the tangent vector on the small scales. Within this picture one can therefore say that the intermittency bursts are related to strong instabilities which grow along directions in phase space which correspond to the small length scales[1]. This fact is very important for the predictability problem, as we discuss in another contribution to this book.

Let us conclude by stressing the weak points of the model. In Eqs.(2) u_n is a complex variable so that the vectorial structure of Navier-Stokes equations is lost. The sensibility of this crude approximation can be decided only a posteriori by the results. The hope of the approach is that the coherent structures widely observed in turbulent fluids do not affect the very nature of scaling, at least on a qualitative level. Recently, some authors have introduced more elaborate shell models where the vectorial structure is not lost, using a so-called Fourier-Weierstrass decomposition[15, 16]. In their approach the shells are coupled over one link but the incompressibility condition is fulfilled within each shell. However these models have much more degrees of freedom than those presented here, and seem to be useful tools for almost realistic simulations rather than for heuristic purposes.

Bibliography

[1] M.H. Jensen, G. Paladin and A. Vulpiani, Phys. Rev. A **43**, 798 (1991).

[2] M.H. Jensen, G. Paladin and A. Vulpiani, Phys. Rev. A **45**, 7214 (1992).

[3] A.M. Obukhov, Atmos. Oceanic Phys. **7**, 41 (1971); A.M. Obukhov, Atmos. Oceanic Phys. **10**, 127 (1974); V.N. Desnyaski and E.A. Novikov, Prikl. Mat. Mekh. **38**, 507 (1974).

[4] E.B. Gledzer, Sov. Phys. Dokl. **18**, 216 (1973).

[5] E.D. Siggia, Phys. Rev. A **15**, 1730 (1977); E.D. Siggia, Phys. Rev. A **17**, 1166 (1978); R.M. Kerr and E.D. Siggia, J. Stat. Phys. **19**, 543 (1978).

[6] R. Grappin, J. Leorat and A. Pouquet, J. de Physique **47**, 1127 (1986); D. Carati and J.-L. Ottinger, unpublished (1993).

[7] M. Yamada and K. Okhitani, J. Phys. Soc. of Japan **56**, 4210 (1987); M. Yamada and K. Okhitani, Progr. Theo. Phys. **79**, 1265 (1988); M. Yamada and K. Okhitani, Phys. Rev. Lett. **60**, 983 (1988).

[8] K. Okhitani and M. Yamada, Progr. Theo. Phys. **81**, 329 (1989).

[9] R. Ruiz and D.R. Nelson, Phys. Rev. A **23**, 3224 (1981).

[10] G. Paladin and A. Vulpiani, Phys. Rep. **156**, 147 (1987).

[11] F. Anselmet, Y. Gagne, E.J. Hopfinger and R. Antonia, J. Fluid Mech. **140**, 63 (1984).

[12] R. Benzi, G. Parisi and L. Biferale, Physica D**65**, 163 (993).

[13] R. Benzi, L. Biferale, G. Paladin, M. Vergassola and A. Vulpiani, Phys. Rev. Lett. **67**, 2299 (1993).

[14] A. Vincent and M. Meneguzzi, J. Fluid. Mech. **225**, 1 (1991).

[15] J. Eggers and S. Grossmann, Physics Lett. A**156**, 444 (1991); J. Eggers and S. Grossmann, Phys. Fluids A**3**, 1958 (1991); S. Grossmann and D. Lohse, Z. Phys. B **89**, 11 (1992); S. Grossmann and D. Lohse, Phys. Rev. Lett. **67**, 445 (1991).

[16] E. Aurell, P. Frick and V. Shaidurov, submitted to Physica D (1993).

Singularities

(and Turbulence)

T. Dombre
C.N.R.S.-C.R.T.B.T.,
Boîte Postale 166, F-38042, Grenoble cedex 9, France
A. Pumir
C.N.R.S.-I.N.L.N.,
1361, route des Lucioles, F-06560, Valbonne, France

INTRODUCTION

A basic tenet of fluid mechanics is that the equations describing the motion of incompressible fluids, given a smooth initial condition, generate regular solutions at all time. Surprisingly enough, this simple proposition is by no means established, and the question of existence of singularities remains unsolved. The problem is more than a mathematical curiosity. One of the most remarkable aspects of fluid turbulence is the occurrence of large excursions of the measured signals (intermittency), suggesting the existence of solutions becoming infinitely large in a finite time. In particular, the violent bursts, observed in turbulent boundary layers[1], might be adequately modeled by blowing-up solutions of the 3-dimensional fluid equations. A variety of presumptions, based on statistical descriptions of turbulent flows[2] or more straightforward power counting arguments, suggests the existence of singularities. To understand in simple terms the (possible) origin of a finite time singularity, let us simply write the 3-dimensional Euler equations in vorticity form :

$$\partial_t \vec{\omega}_i + (\vec{u} \cdot \nabla)\vec{\omega}_i = s_{ij}\omega_j \qquad (1)$$

where ω_i is the ith component of the curl of the velocity, $\vec{u}$, and s is the rate of strain tensor, $s_{ij} = (\partial_i u_j + \partial_j u_i)/2$. Clearly, s and ω are both first order space derivatives of the velocity field. Ignoring the intricate relation between the two quantities, and simply counting powers, Eq.(1) suggests that the norm of vorticity should grow like :

$$\partial_t \Omega \sim \Omega^2 \qquad (2)$$

thus leading to a vorticity blowing-up like $1/(t_* - t)$. The matter is obviously more complicated since the traceless, symmetric tensor s has 3 eigenvalues, associated with 3 orthogonal eigenvectors. Amplification of vorticity is possible only when vorticity is aligned with an eigenvector corresponding to a positive eigenvalue. Also, the effective strain acting upon the vorticity must also become infinite, in order to generate blow-up. In 2 dimensions, the right hand side of Eq.(2) is identically zero, so vorticity is conserved, and smooth solutions remain smooth. These considerations show that the existence of finite time singularities rests on non trivial properties of the equations, and that naive power counting may be deceiving.

A variety of approaches have been proposed to attack the question of existence of finite time singularities. The efforts of mathematicians, physicists and numericists have contributed to sharpen the question, but the final answer is still lacking.

Proving rigorous results for the 3-dimensional, incompressible fluid equations is a delicate problem. Even for the 3D Navier-Stokes equations, generally believed to lead to a well posed problem, incomplete results only are currently avail-

Turbulence: A Tentative Dictionary
Edited by P. Tabeling and O. Cardoso, Plenum Press, New York, 1995

able, so the question of existence of singular solutions remains open. The most interesting mathematical results in our view show that singularities must necessarily have some experimental relevance.

A singular solution to the 3-dimensional Navier-Stokes equations must be such that near the hypothetical singular time t_*, the velocity diverges according to[3]:

$$\|\vec{u}\|_\infty(t) \equiv max_x\|\vec{u}(\vec{x},t)\|$$
$$\geq cst\sqrt{\frac{\nu}{(t_* - t)}} \qquad (3)$$

where the constant is a pure number. Vorticity must also diverge, according to[4]:

$$\|\omega\|_\infty \geq \frac{1}{(t_* - t)}. \qquad (4)$$

From these results, it is possible to give bounds on the amount of energy being dissipated near a singularity, and also on the fractal dimension of the (space-time) fractal set where the solution could be singular[5].

In the inviscid case, a solution becoming singular at t_* must also generate large vorticity, so that[6]:

$$\int_0^{t_*} \|\vec{\omega}\|_\infty(t')dt' = \infty, \qquad (5)$$

implying that vorticity must diverge at least like $1/(t_* - t)$.

A combination of numerical analysis and approximate analytic techniques can also been used to gain some insight on the problem.

Since viscosity is expected to play a regularizing role, the Euler equations are presumably more singular than the Navier-Stokes equations. In the following, we will restrict ourselves to the 3-dimensional Euler equations. We will also consider only solutions with a finite energy, and an initially bounded velocity field. In the next section, we review what has been learned from recent computer experiments of the full Euler equations. In view of the complexity of the full 3-dimensional problem, an interesting idea is to consider some simpler 2-dimensional problem. The most obvious example is the axisymmetric Euler equation. Results in these directions are discussed in the section *RELATED TWO-DIMENSIONAL PROBLEMS* (see further). We conclude in last section.

NUMERICAL STUDY OF THE 3-DIMENSIONAL EULER EQUATIONS

G.I. Taylor and A.E. Green[7] first proposed to consider an initial condition made of a few Fourier modes. The intuition here is that the nonlinearity of the Euler equations will tend to generate higher and higher harmonics as a function of time. The quantitative question is to understand how fast high harmonics grow. To study this question, one can, in principle, generate the coefficients of a Taylor series expansion in time of the Fourier coefficients of the velocity. The existence of a singularity can be deduced from these Taylor series. Unfortunately, the standard methods of extrapolation have proven unreliable for this problem[8].

Instead of considering the Taylor series of $\vec{u}$, it is possible to integrate the initial value problem, using standard pseudo-spectral methods. Several diagnoses have been proposed to track down the existence of a finite time singularity. One is to monitor the growth of the maximum vorticity, since it must become infinite when a singularity develops (Eq.(5)). Another method is based on the concept of analyticity strip. In general smooth functions on the real domain have singularities in the complex plane. Finite time singularities happen when a singularity in the complex plane hits the real domain in a finite time. A function, smooth in a strip of width δ around the real axis, has a Fourier transform that decreases for large wavenumbers faster than $exp(-\delta|k|)$. The spectrum of the solution thus provides information on its analytic structure. In general, singularities tend to approach the real axis. A δ hitting zero in a finite time is a clear sign of singularity formation[9].

Numerical integrations of the Euler equations, with either the Taylor-Green initial condition[8, 10], or with an initial condition made of any set of random modes suggest[10] that δ goes to zero no faster than exponentially in time. Consistent with these results is the fact that the maximum vorticity does not grow very much and that it seems to grow exponentially in time.

In physical space, visualizations of the flow show that vorticity tends to concentrate on large, isolated, vortex sheets, that become thinner and thinner. Elementary fluid dynamics considerations allow to explain why the width of the sheets should go to zero exponentially[10].

An alternative approach to the problem of singularity formation rests on vortex dynamics. Because of the theorems mentioned above, a singular solution to the fluid equations must be associated with a strong self-amplification (stretching) of its vorticity. The problem is therefore to understand in a quantitative manner how vorticity stretches itself.

The simplest objects to study are vortex tubes. The evolution equation for a vortex tube is the well known Biot-Savart equation[11]:

$$\partial_t \vec{r}(\lambda, t) =$$
$$\frac{\Gamma}{4\pi} \int \frac{(\vec{r}(\lambda, t) - \vec{r}(\lambda', t)) \wedge \partial_\lambda \vec{r}(\lambda', t) d\lambda'}{((\vec{r}(\lambda, t) - \vec{r}(\lambda', t))^2 + \sigma^2)^{3/2}} \quad (6)$$

where $\vec{r}$ is the location of the vortex tube, Γ its circulation, and λ a Lagrangian parameter. The core size, σ, is necessary to regularize a logarithmic divergence of the integral. It is the weak point of the approach since the evolution equation for the core size is difficult to derive, and further approximations, such as a fixed core structure, are required to proceed.

Chorin[12] first showed that a coarse version of the Biot-Savart equations can lead to complicated (fractal) structures, suggestive of a finite time singularity. A more quantitative analysis by Siggia[13] showed that generically, a very important stretching happens when two antiparallel pieces of filaments pair. Later, Siggia and Pumir[14] studied systematically the resulting collapse, and showed that in this model, the maximum vorticity becomes infinite, according to $|\vec{\omega}|_\infty \sim 1/(t_* - t)$.

The most questionable aspect of the Biot-Savart model, namely the core structure, has been since thoroughly investigated. Numerical simulations of the Euler equations show that vortex core deformation is very dramatic[15, 16]. The initially antiparallel tubes flatten out very fast, so as to make the situation quasi bidimensional. Since there is no stretching in 2 dimensions, the strain acting on the vorticity comes from remote sources, and remains finite. Vorticity then grows exponentially. The question is not settled, since Kerr[17] found an initial condition where the flow manages to remain genuinely 3-dimensional in the singular region, so the strain does not saturate. This very interesting solution is not fully understood, and calls for more work.

RELATED TWO-DIMENSIONAL PROBLEMS

A fully 3-dimensional velocity field is intrinsically more difficult to study than a 2-dimensional one. This remark has led Grauer and Sideris[18] to study the question of singularity formation in an axisymmetric Euler flow. Denoting by r the distance to the axis of symmetry, z the coordinate along the axis, and ϕ the azimuthal angle, an axisymmetric velocity field is defined by $\partial_\phi \vec{u} = 0$. The Euler equations become in this case:

$$\partial_t u_\phi + (\vec{u}_\parallel \cdot \nabla_\parallel)u_\phi = 0 \quad (7)$$

$$\partial_t \left(\frac{\omega_\phi}{r}\right) + (\vec{u}_\parallel \cdot \nabla_\parallel)\left(\frac{\omega_\phi}{r}\right) = \frac{1}{r^4}\partial_z(ru_\phi)^2 \quad (8)$$

where ω_ϕ is the azimuthal component of vorticity, $\vec{u}_\parallel = u_r \hat{e}_r + u_z \hat{e}_z$ is the velocity vector in a plane $\phi = cst.$, and $\nabla_\parallel = \hat{e}_r \partial_r + \hat{e}_z \partial_z$. Incompressibility implies the existence of a streamfunction, ψ, such that: $u_r = -(\partial_z \psi)/r$ and $u_z = (\partial_r \psi)/r$, satisfying the Poisson equation:

$$(\partial_r^2 + \frac{1}{r}\partial_r + \partial_z^2)\psi = -r\omega_\phi \quad (9)$$

A singular solution away from the origin must generate strong gradients, so that $\partial_r^2 >> 1/r\partial_r$. By systematically neglecting the lowest derivative terms, one is led to the 2-dimensional Boussinesq equations[19, 20]. The azimuthal velocity squared plays the role of temperature in the axisymmetric problem, and centripetal acceleration plays the role of buoyancy. The Boussinesq problem has been proposed as an easier alternative to the axisymmetric problem[19]. Convection in a porous media, governed by Darcy's law, is yet another 2-dimensional, thermally driven flow problem in the same spirit[20].

A similar scenario has been found in both cases. First, the hot parts of the fluid organize into ascending thermal plumes with the largest gradients and vorticity at their leading edges. The front separating the hot and cold fluids gets thinner and thinner thanks to stretching, but keeps its large scale shape for a while. This process leads to an exponential growth of vorticity. Interestingly, the shape of the iso-θ lines eventually becomes wrinkled at smaller and smaller scales. Because of these foldings, the local thickness and radius of curvature of the front remain on the average of the same order of magnitude, allowing thereby finite time blow-up of vorticity. These singularities occur at various places and times along the interface. They are well described by the following scaling laws (with $\tau \equiv t_* - t$)

$$\theta \ \sim \ \tau^n f\left(\frac{\vec{x}}{\tau^{\alpha+\eta}}, -ln(\tau)\right),$$

$$\vec{v} \ \sim \ \tau^{n+\alpha-1}\vec{h}\left(\frac{\vec{x}}{\tau^{\alpha+\eta}}, -ln(\tau)\right), \ (10)$$

compatible with the equations of motion ($\alpha = 2$ for Boussinesq convection, $\alpha = 1$ in porous media and η is much smaller). The presence of the $-ln(\tau)$ argument means that the dynamics remains non trivial during the collapse.

The numerical calculations of Ref. [19] have recently been criticized by E and Shu[21], who argue that numerical noise induces the destabilization of the fronts, as

it is the case in other problems of interface dynamics. This criticism has to be thoroughly answered before definite conclusions can be drawn.

A possible approach beyond numerical analysis consists in continuing the equation of motion off the real domain, and in following the evolution of singularities in the complex plane. This method has been used successfully by Caflisch et al.[22] to study the development of singular solutions of a vortex sheet in 2 dimensions (see also Ref. [23]). Recently Caflisch[24], using an approximation suggested by Moore[25], found singularities in the axisymmetric Euler equations. They seem to correspond more to incipient roll-up of vortex sheets than to the hierarchy of folding instabilities observed numerically. In order to understand the "cascade" of instabilities, Dombre et al.[26] have attempted to capture the proliferation of complex singularities in the early stages of the dynamics (at $t = 0^+$), and to follow their evolution. The structure observed numerically in Ref. [20] suggests a nested set of singularities, with a whole hierarchy of scales, which must hit the real axis in a non trivial manner. We describe here some preliminary results, obtained for porous media convection. It is possible to recast the equation of motion into a contour dynamics form for each iso-θ line, by using the Biot-Savart law. The equations are nonlocal and as such not very transparent. It is useful to continue them analytically in the complex plane, so as to track down the analytic structure of the iso-θ lines. The physical contour gets curved when one of these singularities comes close enough to it. Describing analytically the fate of singularities is easier than the evolution of the interface (see Ref. [26] for the limiting case of a step-like thermal profile). We consider a thermal plume with an initially parabolic shape and a steep but continuous profile. At time $t = 0$, each iso-θ line has only one singularity, all different, but close to one another. This "disordered" structure rapidly collapses into a single singularity, because the buoyancy force makes "hot" singularities move faster than "cold" ones. New singularities are formed in the process on the other side of the physical axis, with very small amplitudes, and close to the symmetry axis of the parabola. These objects could induce at later times folds of small scales, close to the tip of the plume. The process would then repeat for ever, in the absence of thermal diffusion fixing a natural cut-off length.

CONCLUSIONS

We hope to have shown that the problem of finite time singularities in the 3D, incompressible fluid equations remains a major challenge, and as such, a source of inspiration for research. Although the essential question is not yet settled, the efforts over the last years have led to a better understanding of several fluid dynamical issues.

It is a pleasure to acknowledge our collaborators in this work, Eric Siggia and Boris Shraiman.

Bibliography

[1] S. Kline, W. Reynolds, F. Shraub and P. Rundstadler, J. Fluid Mech. **30**, 741 (1967).

[2] M. Lesieur, Turbulence in Fluids, Kluwer, Dordrecht (1990).

[3] J. Leray, Acta Math. **63**, 193 (1934).

[4] L. Caffarelli, R. Kohn and L. Nirenberg, Comm. Pure Appl. Math. **35**, 771 (1982).

[5] V. Scheffer, Commun. Math. Phys. **61**, 41 (1978).

[6] T. Beale, T. Kato and A. Majda, Comm. Math. Phys. **94**, 61 (1984).

[7] G.I. Taylor and A.E. Green, Proc. R. Soc. London Ser. A **158**, 499 (1937).

[8] M. Brachet, D. Meiron, S. Orszag, B. Nickel and U. Frisch, J. Fluid Mech. **130**, 411 (1983).

[9] C. Sulem, P.L. Sulem and H. Frisch, J. Comput. Phys. **50**, 138 (1983).

[10] M. Brachet, M. Meneguzzi, A. Vincent, H. Politano and P.L. Sulem, Phys. Fluids A **4**, 2845 (1992).

[11] D. Moore and P. Saffman, Philos. Trans. R. Soc. London Ser. A **272**, 403 (1972).

[12] A. Chorin, Comm. Math Phys. **83**, 517 (1982).

[13] E. Siggia, Phys. Fluids **28**, 794 (1985).

[14] E. Siggia and A. Pumir, Phys. Rev. Lett. **55**, 1749 (1985).

[15] A. Pumir and E. Siggia, Phys. Fluids **A2**, 220 (1990).

[16] M. Shelley, D. Meiron and S. Orszag, J. Fluid Mech. **246**, 613 (1993).

[17] R. Kerr, Phys. Fluids A **5**, 1725 (1993).

[18] R. Grauer and T. Sideris, Phys. Rev. Lett. **67**, 3511 (1991).

[19] A. Pumir and E. Siggia, Phys. Fluids A **4**, 1472 (1992).

[20] A. Pumir, B. Shraiman and E. Siggia, Phys. Rev. A **45**, 5351 (1992).

[21] W. E and C. Shu, Phys. Fluids, **6**, 49 (1994).

[22] R. Caflisch, N. Ercolani and T. Hou, Comm. Pure Appl. Math. **46**, 453 (1993).

[23] S. Tanveer, Phys. Fluids A **5**, 1456 (1993).

[24] R. Caflisch, Physica D **67**, 1 (1993).

[25] D. Moore, Proc. R. Soc. London Ser. A **365**, 105 (1979).

[26] T. Dombre, A. Pumir and E. Siggia, Physica D **57**, 311 (1992).

SPATIOTEMPORAL INTERMITTENCY

H. CHATÉ and P. MANNEVILLE
CEA, Service de Physique de l'Etat Condensé, Centre d'Etudes de Saclay,
F-91191 Gif-sur-Yvette, France
LadHyX — Laboratoire d'Hydrodynamique, Ecole Polytechnique,
F-91128 Palaiseau, France

SPATIOTEMPORAL INTERMITTENCY: although not to be found in a standard English dictionary, this expression can easily be understood at an "ordinary language" level and indeed appears in the physics literature with this general, and hence vague, meaning. However, spatiotemporal intermittency has recently been used to define dynamical regimes possessing a precise set of properties and involving specific physics. In the following, these two levels of description are analyzed successively and the theoretical implications of the latter are discussed.

SPATIOTEMPORAL INTERMITTENCY IN A LOOSE SENSE

In loosely defined language, spatiotemporal intermittency is used to describe the evolution of a (dynamical) system for which the space-time representation of one of its observable has an intermittent character. Trivial limits are excluded from this definition. This is the case of zero-dimensional systems, such as the chaotic dynamical systems describing the purely temporal disorder of some physical situations, since they have no spatial extension. Also excluded are spatially extended systems for which the disorder is either purely temporal —since their frozen spatial structure allows a reduction to the zero-dimensional case above— or purely spatial —since they have no dynamics and fall into the category of static disorder problems.

The intermittent character referred to above assumes that there is disorder in the system. It has to be understood, not in the rather well-defined meaning it has been given in the context of the "universal routes to chaos"[1], but rather in the looser sense sometimes used in fluid turbulence when, e.g. the irregular passage of turbulent plugs in a pipe flow is represented by a telegraphic signal.

Spatiotemporal intermittency thus refers to situations in which both snapshots of this observable in the system and its local time series are akin to telegraphic signals. Defined this way, intermittency assumes the existence of *two local states* in the system (related, e.g., to the two-hump pdf of some local variable).

SPATIOTEMPORAL INTERMITTENCY IN A STRICT SENSE

The above definition makes no assumptions about the processes governing the dynamics of the local states and especially the evolution of the borders between spatial domains of different states. In most situations of interest, one finds a regular and an irregular local state, or, to follow up with the pipe flow image, a "laminar" and a "turbulent" local state. The stricter definition of spatiotemporal intermittency limits the above "loose" definition to situations where only two local states are observed, one regular/laminar and one irregular/turbulent. The crucial additional property required is that the regular/laminar state be *absorbing*, i.e.

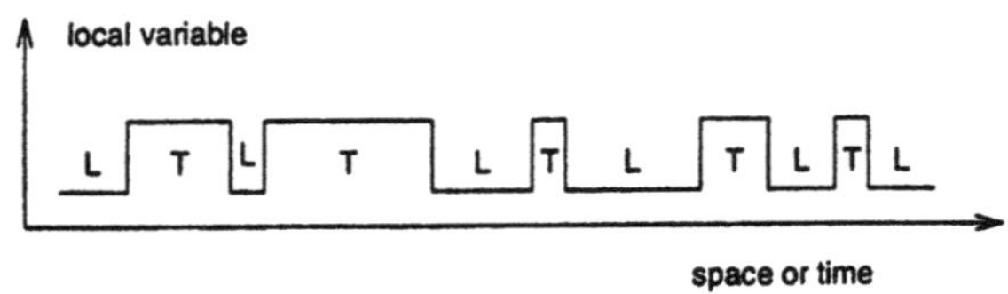

Figure 1: Schematic representation of an intermittent "telegraphic" signal. Symbol "L" denotes the regular/laminar local state, symbol "T" the irregular/turbulent one.

that the irregular/turbulent state does not appear spontaneously in the middle of a regular region, and that the interplay between the two local states is governed by a *local* recession/invasion process at the borders.

The schematic spatiotemporal evolution represented in fig. 2a does not fit into this stricter definition of spatiotemporal intermittency. At two points in space-time, a laminar region breaks down in the middle to create a turbulent domain. In fig. 2b, on the other hand, no such events are observed, and the contaminative nature of the propagation of the irregular/turbulent state is obvious. Spatiotemporal intermittency thus implies a strong and essential asymmetry between the two local states: whereas the irregular/turbulent state can spontaneously decay to the regular/laminar state (due to its intrinsic fluctuations), the inverse transition does not occur (the laminar/regular state has no or too weak intrinsic fluctuations) and can only be triggered by neighboring turbulent domains.

The above definition can, in fact, be stretched to include situations with more than two local states, provided that one of them is absorbing.

EXPERIMENTAL RELEVANCE AND THEORETICAL MOTIVATION

Recently, a number of carefully controlled laboratory experiments have been performed which produced spatiotemporal intermittency regimes. Examples can be found in the article "Experiments in 1D turbulence" by F. Daviaud in this volume. In fact, the above definition leaves

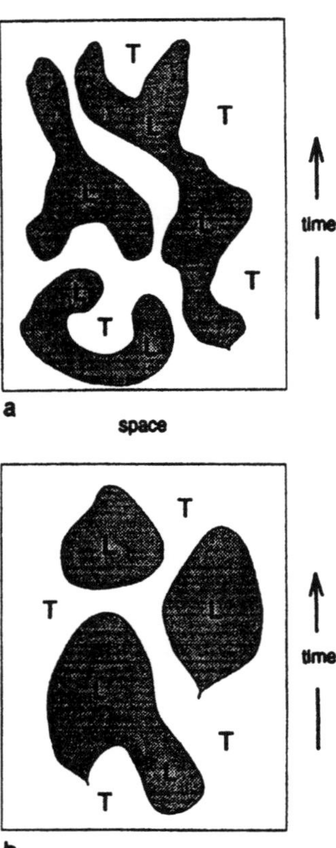

Figure 2: Schematic representation of spatiotemporal intermittency regimes. The regular/laminar local state is grey (symbol "L"), the irregular/turbulent one white (symbol "T"). In part (a), there is no asymmetry between the two local states (STI in a loose sense), whereas in part (b) the regular/laminar state is absorbing (STI in a strict sense).

room for many other experimental situations to qualify as spatiotemporal intermittency. In particular, electrohydrodynamic convection in nematic liquid crystals[10], capillary ripples in the Faraday instability, and directional solidification in eutectics are but some of the physical problems in which spatiotemporal intermittency can also be observed but they have not yet been studied as such. In spite of the diversity of these situations, one can study them within a common theoretical framework, as originally suggested by Y. Pomeau[3]. As a matter of fact, the requirements for spatiotemporal intermittency (extended system, local interactions, two local states with one being absorbing) constitute the essential ingredients of *directed percolation*, a well-known probabilistic cellular automaton modeling various contamination phenomena[11].

Directed percolation, when seen as a dynamical process, iterates synchronously, at discrete timesteps, the sites of a lattice according to transition probabilities which depend on the configuration of their neighborhood. Each site is in one of two possible states, an "active" state and an absorbing state. The transition probabilities governing the evolution of the parental configurations generally depend on control parameters except for the configuration in which all the sites are in the absorbing state: in this case, the central site has to remain in the absorbing state at the next timestep. A simple case is directed *bond* percolation in one space dimension for which there is a unique control parameter p (the probability for a bond to be "opened") in terms of which the various transition probabilities are expressed. Varying p, a continuous (second-order-like) phase transition occurs: at low p, any initial condition eventually reaches a fully absorbing homogeneous state. Above a critical value p_c, the active state percolates to infinity: there are sites in the active state at any given time.

The analogy with spatiotemporal intermittency should be clear by now: the absorbing state corresponds to the regular/laminar state, the active state to the irregular/turbulent one. One of the main interests of this analogy is its suggestion that the transition from fully regular/laminar situations to spatiotemporal intermittency regimes might be of the same nature as the directed percolation transition, and, in particular, that they could be governed the same "universal" critical exponents. For example, the mean turbulent fraction in the spatiotemporal intermittency regimes could go continuously to zero when approaching the threshold, as does the density d of active sites in directed bond percolation: $d \sim (p - p_c)^\beta$ with $\beta \simeq 0.28$ in one space dimension. At any rate, the analogy calls for a description of the transition to turbulence via spatiotemporal intermittency in terms of a *phase transition*, suggesting a thermodynamics whose building blocks are not atoms or molecules but macroscopic units arising from some instability mechanism.

ANALYZING SPATIOTEMPORAL DATA: A TYPICAL CASE

The analogy proposed with directed percolation also provides guidelines for analyzing the mass of spatiotemporal data produced by experiments of spatiotemporal intermittency, be they numerical or not. The first step consists, of course, in reducing the experimental data to probabilistic cellular automata-like ones, i.e. to perform a binary reduction of the local variables based on the two local states, possibly followed by discretization of space and time.

Given the wide variety of systems in which spatiotemporal intermittency can be observed, there is no "universal", systematic reduction method. Except in cases where the two local states can be exactly defined (see the example below), one usually needs to introduce some "artificial" thresholding of the data, as is often done in image-processing techniques. The independence of the statistical results on the various cut-offs introduced must then be checked. Once the reduction method is fixed, one can measure the quantities used to characterize directed percolation regimes. This is illustrated in the following example.

As mentioned above, one of the crucial theoretical points raised by the analogy with directed percolation is that of the "universality" of the critical exponents possibly governing the transition region. Continuous transitions like the directed percolation transition are accompanied by important finite-size effects which have to be assessed to properly evaluate the asymptotic values of the exponents. This usually requires observation of systems of large size. This is the main reason that a "minimal" coupled map lattice model, possessing the essential properties of spatiotemporal intermittency while being as simple as possible, was introduced[4]. All the sites of this numerically-efficient, discrete-space, discrete-time dynamical system are updated synchronously according to the following (here in one space dimension):

$$X_i^{t+1} = f(X_i^t) + \frac{\varepsilon}{2}(f(X_{i-1}^t) - 2f(X_i^t) + f(X_{i+1}^t))$$

where t is the discrete time, X_i is a real variable sitting at site i, $f(X)$ is the local map, and ε is the strength of the diffusive coupling between a site and its nearest neighbors. The piece-wise linear map f is defined by $f(X) = rX$ if $x < 1/2$, $f(X) = r(1 - X)$ if $1/2 < X < 1$ and $f(X) = X$ if $X > 1$. It is composed of a chaotic repellor ($X < 1$) connected to a continuum of marginally stable fixed points ($X > 1$). The sought-after binary reduction is naturally based on this separation of the local phase space of the maps. At time t, site i is in the regular/laminar (respectively irregular/turbulent) state if $X_i^t > 1$ (respectively $X_i^t < 1$). It is easy to show that the regular state is absorbing, thanks to the smoothing nature of the coupling.

This model shows spatiotemporal intermittency: whereas, for small ε, any initial condition will eventually lead to a fully absorbing state ($\forall i, \quad X_i > 1$), there exists a threshold value ε_c, well-defined in the infinite-size limit, above which disordered initial conditions lead to statistically stationary regimes of spatiotemporal intermittency (fig. 3). Following directed percolation studies, various quantities can be measured after the binary reduction has been performed. For example:

- the variation of the mean turbulent fraction f_{turb}, i.e. the concentration of sites in the irregular local state, with the coupling strength ε. In the present case, $f_{turb} \sim (\varepsilon - \varepsilon_c)^\beta$.

- at a given $\varepsilon \geq \varepsilon_c$, the distribution of sizes and lifetimes of the regular/laminar domains. Above threshold they are usually distributed exponentially (allowing the definition of coherence scales $\xi_\parallel$ and $\xi_\perp$), while at threshold they might be distributed algebraically as in the present case.

- the variation with ε of the above-defined scales $\xi_\parallel$ and $\xi_\perp$. They might diverge at threshold, like $\xi_{\parallel/\perp} \sim (\varepsilon - \varepsilon_c)^{-\nu_{\parallel/\perp}}$.

- the mean duration τ, for $\varepsilon < \varepsilon_c$, of the transients leading to the completely laminar state. It might diverge around threshold, like $\tau \sim (\varepsilon_c - \varepsilon)^z$.

For all these quantities, the finite-size effects have to be assessed, leading, in the case of a continuous transition like that presented here, to finite-size scaling analysis. In laboratory experiments and, to a lesser extent, in numerical simulations, this is sometimes impossible; so that when a continuous (second-order-like) transition to spatiotemporal intermittency is found, the measured values of the critical exponents are not the asymptotic ones characterizing the system in the thermodynamic limit. Moreover, only some of the above quantities may be accessible, due to the limitations of the data acquisition technique and/or the quality of the data themselves. Such a situation does not

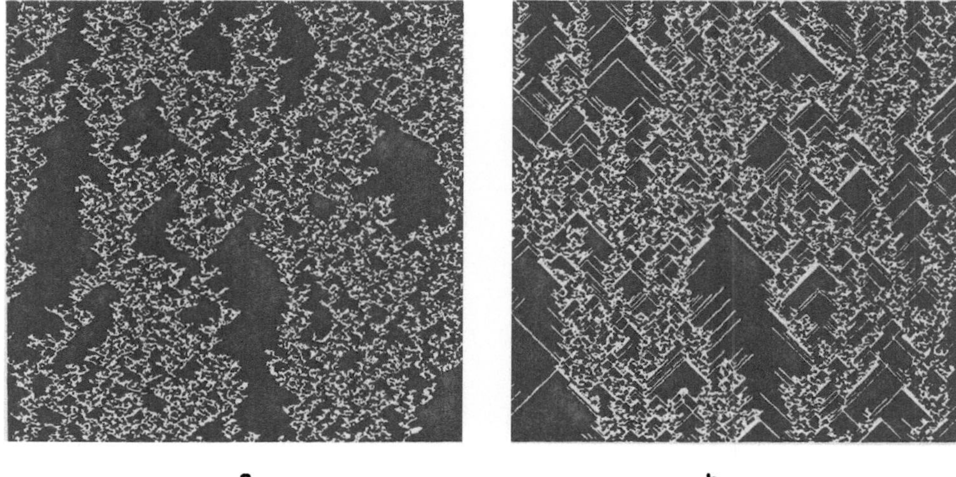

Figure 3: Spatiotemporal intermittency regimes of the minimal coupled map lattice model defined in the text. The build-in binary reduction is used: laminar/regular sites are in black, turbulent/irregular ones in white. Spatiotemporal representation with time running upward. Lattice of $N = 512$ sites with periodic boundary conditions. Transients following random initial conditions were discarded. Part (a): $r = 2.1$, $\varepsilon = 0.0048$, 512×16 iterations. Part (b): $r = 3$, $\varepsilon = 0.361$, 512 iterations.

mean that the transition to spatiotemporal intermittency cannot be properly studied. As a matter of fact, thanks to the (necessary and observed) ergodicity in space-time of spatiotemporal intermittency regimes, even a *local* time series of one variable may be enough, although the acquisition of data over very long periods of time in order to compile satisfactory statistics. No miracle should yet be expected: in all cases, a truly large, extended system is needed to observe spatiotemporal intermittency. Intermittent local time series from a confined system have more to do with low-dimensional chaos[1].

NON-UNIVERSALITY OF THE TRANSITION TO SPATIOTEMPORAL INTERMITTENCY

The minimal model presented above, studied in one and two space dimensions, has shown that the transition to spatiotemporal intermittency does *not* possess the universal character suggested by the analogy with directed percolation[4]. The most striking departure from directed percolation is observed when the local map f is slightly modified so that the regular/laminar local state possesses a stable fixed point (e.g. $f(X) = kX + (1 - k)X^*$ with $|k| < 1$ and $X^* = (r + 2)/4$ for $X > 1$). In this case, the transition to spatiotemporal intermittency is *discontinuous* (first-order-like) and no critical region is present. But even in the case where the transition is continuous, the critical exponents measured are not those of directed percolation and vary with the details of the model (for example with the parameter r in the model above). Of course, in the absence of a theory, nothing thus far rules out the possibility that the observed scaling is only intermediate and that the scaling of directed percolation could be recovered at scales out of reach of current numerical power[7]. Also at odds with directed percolation is the apparent breakdown of the hyperscaling relationship linking the critical exponents[6].

This somewhat disheartening conclusion must be tempered. All the models and experiments in which spatiotemporal intermittency has been observed suggest that roughly two types of regimes can be distinguished, calling for two different theoretical approaches[5]. In the first type, the spatiotemporal disorder arises from the interaction of localized, propagative, soliton-like objects clearly related to the deterministic nature of the system (fig. 3b). The second type is closer to directed percolation. In fig. 3a, no obvious "deterministic" objects are visible. Rather, as in directed percolation, the propagation of disorder seems to be governed by "effective" probabilistic rules arising from the strong local chaotic mixing processes in the system.

Two directions have thus to be considered for further theoretical progress. On the one hand, the relevant localized objects at the origin of the spatiotemporal dynamics can be studied, in the hope of building some "kinetic theory". On the other hand, one can try to estimate the local transition rules in order to construct an equivalent probabilistic cellular automaton.

Bibliography

- A general introduction to **chaos** (including the route to chaos via intermittency mentioned above), as well as an account of current knowledge on spatially extended situations can be found in:

[1] P. Manneville, *Dissipative Structures and Weak Turbulence*, Academic Press, New-York, 1990.

- Articles on **spatiotemporal intermittency** tend to be numerous, but only part of them actually deal with the process defined here (spatiotemporal intermittency in a strict sense). A review is:

[2] H. Chaté, "Subcritical Bifurcations and Spatiotemporal Intermittency." in: *Spontaneous Formation of Space-Time Structures and Criticality*, edited by T. Riste and D. Sherrington, 273-311. Kluwer, Dordrecht, 1991.

- The seminal work of Y. Pomeau is:

[3] Y. Pomeau, *Physica D* **23** (1986) 1.

- Papers on the **minimal coupled map lattice model** introduced above include:

[4] H. Chaté and P. Manneville, *Physica D* **32** (1988) 409; *Europhys. Lett.* **6** (1988) 591; *Physica D* **37** (1989) 33.

[5] H. Chaté and P. Manneville, *J. Stat. Phys.* **56** (1989) 357; *Physica D* **45** (1990) 122.

[6] J.M. Houlrik, I. Webman and M.H. Jensen, *Phys. Rev. A* **41** (1990) 4210; J.M. Houlrik and M.H. Jensen *Phys. Lett. A* **163** (1992) 275.

[7] P. Grassberger and T. Schreiber, *Physica D* **50** (1991) 177.

- For **spatiotemporal intermittency in partial differential equations**, one can look at:

[8] H. Chaté and P. Manneville, *Phys. Rev. Lett.* **58** (1987) 112.

[9] H. Chaté, "Spatiotemporal intermittency in the one-dimensional complex Ginzburg-Landau equation", *Nonlinearity*, **7** (1994) 185

- **Laboratory experiments** showing spatiotemporal intermittency are described in the article "Experiments in 1D turbulence" by F. Daviaud in this volume where original references can be found. Recently, S. Nasuno has started looking at the breakdown of the grid pattern in nematic liquid crystals in terms of spatiotemporal intermittency in two space dimensions.

[10] S. Nasuno, private communication.

- **Directed percolation**, its critical exponents, finite-size scaling and hyperscaling relation are well described in:

[11] W. Kinzel, in *Percolation structures and processes, Ann. Israel Phys. Soc.* **5** (1983) 425.

STATISTICAL APPROACH
(to 2D turbulence)

Y. POMEAU

Laboratoire de Physique Statistique, associé au CNRS,
Ecole Normale Supérieure
24, rue Lhomond, F-75231 Paris cedex 05, France

I insist mostly on the most recent theoretical results, that give a consistent, and probably true overall picture. I explain also briefly how some novel statistical idea on kinetic theory with inelastic interactions may be used to understand the decay laws of population of patches of vorticity.

The application of statistical ideas to turbulence remains a notoriously difficult subject. To start from the very beginning, it is of interest to understand why a straight application of equilibrium Gibbs-Boltzmann statistics fails. This is because of the so-called Jeans paradox. The dynamics of inviscid flows is a particular example of classical field theory with a Hamiltonian structure. For such a system, each degree of freedom at equilibrium has on average the same amount of energy, and as there are infinitely many degrees of freedom in a field (the short wavelength fluctuations), everything is divergent at thermal equilibrium, unless something very special happens (as it does in 2D Euler flows). Whence the Kolmogorov idea of a cascade: energy is injected at large scales and dissipated at small scales. Presently, in 3D flows the statistical theory behind the cascade picture is still rather blurred, although in 2D important progress has been made recently, that will make the matter of this entry in the "dictionary of turbulence".

The focus will be mostly on what followed the fundamental paper by Onsager on this question[1]. This line of approach carefully avoids discussing matters as a Kolmogorov spectrum in 2D, since it puts the emphasis mostly on large scale structures that are precisely beyond reach of spectral theories. Such spectral theories do not deal with the kind of information necessary for understanding coherent phenomena, as they assume more or less a random phase approximation, that kills any spatial flow structure. Reference to the spectral point of view in 2D turbulence may be found in the review by Kraichnan and Montgomery[2].

The first hint to a statistical approach to the 2D turbulence problem was given by Onsager[1]. His idea can be presented as follows: 2D Euler equations form a Hamiltonian system, but with a very special property, as it has infinitely many conservation laws (not to be confused with the infinitely many conservation laws of integrable systems, as 2D Euler is definitely not integrable, the four vortex problem having been shown[3] to have a heteroclinic connection). The conserved quantities are the values of the vorticity at different points in the fluid. The Euler equations for a 2D incompressible flow can be written in an almost infinite number of ways. To emphasize the role of vorticity conservation, on can use the following form:

$$\frac{\partial \omega}{\partial t} + u \nabla \omega = 0, \qquad (1)$$

where ω is the vorticity (a pseudoscalar field in 2D), and u is the incompressible velocity field, with two Cartesian components u and v, both functions of x and y. Incompressibility in 2D is enforced by

Turbulence: A Tentative Dictionary
Edited by P. Tabeling and O. Cardoso, Plenum Press, New York, 1995

introducing the stream function $\Psi(x, y; t)$ such that

$$u = -\frac{\partial \Psi}{\partial y} \quad \text{and} \quad v = \frac{\partial \Psi}{\partial x};$$

the vorticity is now $\omega = -\nabla^2 \Psi$. Periodic boundary conditions will be assumed most of the time. From (1), the vorticity is conserved in the sense that the area in the fluid where the vorticity has a given value remains constant in the course of time. Whence the infinite number of conserved quantities: if the vorticity has infinitely many values, the histogram of these values weighted with the covered area is constant in the course of time. It is indeed a nontrivial task to put this to work in the framework of Gibbs-Boltzmann statistics. Gibbs-Boltzmann statistics is based upon the ergodic assumption, and the statistical ensemble for the fluctuations depend explicitly on the conserved quantities. This means that, in this specific case of 2D Euler system, infinitely many conserved quantities should be included into the form of the statistical ensemble, a nontrivial task that I shall consider later on. Onsager simplified the matter by assuming vorticity to be concentrated into finitely many point vortices. The dynamics of point vortices has been long a subject of investigations. Famous in particular is the attempt by Kelvin (Thomson, later dubbed Lord Kelvin) to describe atoms as equilibrium configurations of vortices, not such a bad idea in this time. A thorough account of point vortex dynamics may be found in R. Comte Ph.D. thesis[4] as well as in the articles by Aref[5] and Aref and Pomphrey[6]. The point of view of Onsager was slightly different, as he used only the property that the equations of this system of point vortices are Hamiltonian, which allows at once to write the probability distribution of a configuration of vortices as

$$\frac{1}{Z} exp(-\beta H),$$

where Z is the normalization factor (or partition function), and β the inverse temperature. This temperature is related to the Lagrange multiplier for the energy. The Hamiltonian H, a function of the position of the vortices in the system, reads:

$$H = \sum_{j \neq i} \Gamma_i \Gamma_j \ln |r_i - r_j|,$$

where the sum is over all vortices of index j, different from i and where Γ_l is the circulation around the vortex l. If the vortices are to stay in a closed "box", the phase space is compact, and so the temperature can be either positive or negative, as was noticed by Onsager. A negative temperature means that, contrary to what happens in ordinary statistical physics, high energy states are statistically favored with respect to the low energy ones, having a higher probability (for β negative the Boltzmann factor is the biggest when the energy is the largest). These high energy states correspond to the clustering of vortices with the same sign (vortices are like electric charges: like sign vortices increase the total energy when they come close together). Onsager predicted then that, in 2D Euler flows, one should observe the formation of large scale vortices, understood as patches of vorticity with a given sign. The negative temperature state of a system of vortices of zero global circulation should be made of two large vortices, each with a given sign.

The thermodynamics of these point vortices is in itself an interesting field of study, as it is related to other areas of statistical physics through the remark that defects in some 2D field theories may interact like Kelvin's point vortices. The equation of state may be obtained quite simply by noticing that the Gibbs-Boltzmann statistical weight is a product of powers of $|r_i - r_j|$, as it is the exponential of a logarithm. A most remarkable result was obtained by Jancovici[7], who has been able to compute exactly the thermodynamical properties of this system when the power is exactly equal to (-2).

To come back to the fluid mechanical problem, it was appreciated early that the point vortex model cannot explain everything, since it remains to understand how one can translate, if possible, results from point vortex theory to the case of a continuous vorticity. It is fair to say that a continuous distribution of vorticity is not always the good physical model. In superfluids, the vorticity is concentrated on singularities, points in 2D, lines in 3D. However classical hydrodynamics assume a continuous and bounded vorticity (this is the condition needed for proving the regularity of the Cauchy problem in 2D), and the problem remains to extend the Onsager theory to these continuous models. This is a nontrivial endeavor, and a rather complete story of it can be found in the recent paper by Miller *et al.*[8]. In some sense the conservation of vorticity in continuous system is transferred to the explicit introduction of the Γ's in the Hamiltonian, and the limit amounts to replace the discrete set of Γ by a continuous field. This highly singular limit has needed a new idea that I shall explain now.

Equation (1) is in the form of a advection equation, that is ω is like a concentration field of an impurity for instance (the so-called scalar field). Indeed ω is not a passive scalar, because its distribution allows ultimately to determine the velocity field **u** responsible of the advection. Let us assume nevertheless that ω diffuses as a passive scalar, the so-called "Lagrangian" diffusion (a better name would be "Gibbsian", as I am not sure that Lagrange had ever this idea of diffusion by a deterministic velocity field). This Lagrangian diffusion leads to the fact that smaller and smaller length scales appear in the vorticity distribution. Let $\lambda(t)$ be the typical length scale for this distribution, that should tend to zero as time goes on (this is an example of irreversibility in a formally reversible process). Because vorticity is conserved in the course of time, the order of magnitude of the vorticity is constant during this Lagrangian diffusion. Let ω be this order of magnitude, as the vorticity is like a space-derivative of the velocity, the velocity should scale like $u \approx \lambda\omega$, and thus tend to zero for long times. This is not compatible with

the conservation of energy, as the energy per unit area is proportional to u^2. The conservation of energy is realized by the formation of large scale structures, that are not covered by the argument on the cascade to smaller length scales. Note that this argument is in contrast with the one relevant for other classical field theories. For instance it has been shown[9] that the nonlinear Schrödinger equation in the defocusing case and for $D > 2$ presents a cascade toward the small scales, without any formation of large scale structure. This sketch is still far from giving a precise method of computing equilibria of 2D Euler flows with a smooth vorticity. However it was at the basis of the work presented at about the same time by Miller[10] and by Robert and Sommeria[11]. To make a long story short, the idea there is that there are two levels of description of the vorticity field: on large length scales, this field is described by averaging at any point over a local equilibrium of the vorticity in a flow, and this local equilibrium results from the exchange of small scale vorticity between different regions: the maximization of the entropy of mixing is done by introducing an uniform chemical potential that rules the exchange from one part of the system to the other. It is absolutely remarkable that this yields a closed set of equations relating explicitly the final equilibrium state to the value of the conserved quantities in the initial data.

Let us write the equations in the case where the initial vorticity field takes a discrete set of values $\omega_1, \omega_2, ...\omega_i, ...$ each filling an area $\Omega_1, \Omega_2, ...\Omega_i, ...$ (Below I shall use as far as possible notations close to Robert and Sommeria[11]). The short scale fluctuations of the vorticity take place in a nonuniform coarse grained velocity field. Let $\Psi(r)$ be the stream function of this coarse grained field. The relative amount of area covered by the vorticity ω_i near **r** follows from a maximization of the entropy of mixing, constrained by the energy conservation and by an uniform chemical potential α_i, and it is given by a Boltzmann factor:

$$e_i(r) = [Z\{\Psi(r)\}]^{-1}$$
$$\exp[-\alpha_i - \beta\omega_i\Psi(r)]. \quad (2)$$

In this expression, I have introduced the inverse temperature β which will be the Lagrange multiplier for the energy. Since $e_i(r)$ is a proportion, these proportions have to sum to one, which defines the partition function $Z\{\Psi(r)\}$ as:

$$Z\{\Psi(r)\} = \sum_i \exp[-\alpha_i - \beta\omega_i\Psi(r)]. \quad (3)$$

The Laplacian of the coarse grained stream function at any point $\mathbf{r}$ has to be equal to minus the mean value of the vorticity (mean value computed over the local statistical Gibbs ensemble), that gives the relation:

$$-\nabla^2\Psi = \sum_i \omega_i e_i(r). \qquad (4.a)$$

This is to be solved (for Ψ) together with boundary conditions. In the case of rigid boundary condition, this boundary has to be a flow line that can be taken as the zero level line of Ψ:

$$\Psi = 0 \quad \text{on} \quad \partial\Omega. \qquad (4.b)$$

It remains to define now the Lagrange multipliers α_i and β. They are defined by the initial data: the energy and the area covered by each value of the vorticity ω_i. The condition on the area reads:

$$\int_\Omega e_i(r)dr = \Omega_i$$

and for the energy:

$$E = \frac{1}{2}\int_\Omega [\nabla\Psi]^2 dr,$$

where E is the total energy.

The number of conditions equals the number of unknowns. Robert and Sommeria[12] have studied the number of solutions of this system, and the conditions under which it has an unique solution. In particular, when this solution is unique, it is stable in the usual mechanical sense. Moreover, it shows concentrated patches of like sign vorticity when the energy is large enough, which would correspond to the negative temperature state of Onsager.

It is expected that this equilibrium state is reached after a long evolution of the system and that it will stay there afterwards, as a true equilibrium state should do. In the present case however, there is a subtle difference with ordinary statistical mechanics of system of particles, that has to be considered. In many body systems, one expects the ergodic assumption to hold, which, although unproven in general in a mathematical sense, has every chance to be true. This is because the motion of all the individual particles put enough randomness in the system to make it ergodic. For the present problem, the source of randomness is far weaker: it is made of the short range fluctuations of the vorticity, which, as we have seen, become more and more inefficient to generate fluctuations (and then randomness) of the velocity field as times goes on. Moreover, as the final coarse grained velocity field is both stationary and 2D, it certainly has no mixing effect, having closed flow lines. Thus, although the equilibrium theory developed so far is reasonably self consistent, it is not completely obvious that it can be applied to all sorts of situations. The phase space could perhaps be divided in more than one component. This is difficult to formulate more precisely, because one is *a priori* lacking of an obvious invariant Liouville measure for systems with infinitely many degrees of freedom like this one.

A related question is the assumption of uniformity of the chemical potential. If one thinks to usual statistical mechanics, this amounts to suppose that there is a unique way of counting the number of configurations in phase space. This is insured by the general principles of statistical mechanics.

In other terms, everything there is as if there were a unique scale for the smallest piece of phase space. This elementary unit of volume in classical (= nonquantal) phase space is often-and erroneously-presented as being the quantum scale h^{3N}, h Planck's constant and N number of particles. In the present system, there is no obvious justification to the choice of an uniform small scale for counting the number of states, particularly because the final state predicted to show-up is nonuniform. This uniform scale should be there if small scale vorticity is exchanged fast enough between different parts of the system, which is not obvious at all.

Henceforth, it seems of particular interest to look at the evolution toward this "final" state. This was done in two different ways. Robert and Sommeria[13] have devised a partly phenomenological theory for the close approach to equilibrium. I refer to the original paper for this. Motivated by the need to write about statistical theories of 2D turbulence, I will present a theory of the evolution at much earlier times, that uses some new ideas of kinetic theory of systems with inelastic interactions. The inelastic interactions under consideration are the merging of like sign vortices, and they can be put in a dynamical picture for the formation of larger and larger vortices in the course of time that will ultimately end in the form of two large vortices.

As explained in Ref. [15], one can understand the process of formation of these large structures in the limit where there is more vorticity than energy in the initial conditions. Let E be the initial energy in the system, although ω is still the vorticity magnitude. Let furthermore Λ be the size of the system (for instance the period of the periodic box), then one may form the dimensionless ratio $p = \omega^2 \Lambda^4 / E$.

This quantity cannot be small (assuming that the velocity has no potential component — for instance uniform), because the length scale is bounded from above by Λ and because p measures Λ with respect to the typical length scale. On the other hand p can be very large by the argument used before, because when the vorticity varies on short distances the velocity — and then the energy — are relatively small.

One expects that in this limit of p large, the conservation of energy will not be a too strong constraint on the dynamics, as it has to do with the energy that is precisely relatively small. In this limit then, one expects that the coherent structures storing the energy fill a relatively small area of the system and then evolve more or less like point vortices. Nevertheless this is not the whole story, as it has been shown convincingly[14] that when two of these vortices with the same sign pass close to each other, they merge so that the number of vortices decreases in the course of time. Ultimately indeed these vortices will form a single pair, as predicted[10, 11].

The numerical simulations show that the number of vortices, when it is still large, decays according to a power law. If $N(t)$ is this number at time t, this decay law reads:

$$N(t) \approx N_0 \left(\frac{t}{\tau}\right)^{-\xi}$$

where τ is a reference time and ξ an exponent. The value of the exponent so defined seems to be universal and about 0.70-0.75. Looking at the available theories, we found this to be inconsistent with them. Batchelor proposed[16] that the scaling for 2D turbulence is with the energy instead of the energy dissipation in 3D-Kolmogorovian turbulence, because no energy cascade takes place in 2D (if I understand well the argument). This yields $\xi=2$, clearly outside of the error bars. This shows the need of another theory, which is sketched below. One major difficulty with this subject is that the scaling approach does not seem relevant for 2D turbulence, mainly because there are too many conserved quantities to yield unique power laws. Thus it seems important to try first to understand the physics of the process one wants to describe and then to derive exponents for the decay law or at least relations between different exponents.

The first thing one has to understand is what causes the merging of like sign vortices. To make things simpler, let us think in terms of order of magnitude. The dynamics of two vortex patches is constrained by vorticity and energy conservation. Let a be the order of magnitude of the radius of the two vortices. Each one carries an energy $\omega^2 a^4$, neglecting long range logarithmic contributions (see discussion below). The vortices merge, if they have the same sign, in order to minimize the area covered by the organized flow: according to the arguments of Refs. [10,11], the mixing entropy of the vorticity field is proportional to the area covered by the well-mixed part of the vorticity. Thus, to maximize this entropy, it is better to leave as much area as possible outside of the coherent vortices. As a result the vortices merge, because, when doing so, and because the energy must be conserved, the fourth powers of the radius add, and so — by Schwartz inequality — the resulting vortex covers less area than the two vortices before.

This leaves open the fate of the rest of the vorticity that was stored inside the merging vortices and should be dumped in some way or another after the merging, because not enough area is left in the final vortex. This problem is to be compounded with the long range character of the vortex energy. Actually the energy of a vortex in a background of other vortices, opposite and like sign, is the sum of a self-energy and of an interaction part. The interaction part is like $\sum_{j\neq i} \Gamma_i \Gamma_j \ln |r_i - r_j|$, where the sum is over all vortices of index j, different from i and where Γ_l is the circulation around the vortex l. On average the interaction energy has the same scalings as the self energy, if the energy of the system is extensive, and is proportional to the number of vortices, the sum over j in the interaction term together with the logarithm in the self energy leads to a cancellation of any divergence in the thermodynamic

limit, because of the equal number of plus and minus circulation Γ_l (I assume the total circulation to be zero). This shows that if the merging reduces the area covered by the vortices, on average this will change in the same way the self energy and the interaction energy, and so this last one will remain constant if the sum of all self energies is conserved, as insured by adding the fourth power of the radius on the merging event (this assumes implicitly that the ratio of the interaction to the self energy remains constant in the course of time).

A first obvious consequence of these considerations is that the average radius of the vortices grows like $(t/\tau)^{\xi/4}$, in good agreement with the numerics. Similarly, the Kurtosis scales like $(t/\tau)^{-\xi/2}$, as predicted by this theory[15], in agreement too with the numerics.

This presentation leaves two points unsettled: i) Is there a sensible way of calculating the exponent ξ? ii) What can one conclude from this concerning flows kept out of equilibrium by a constant stirring of one sort or another? The second point has not yet received the attention it deserves, although the first one is considered in some details in Ref. [17]. A "reasonable" explanation of the universality of ξ may be found. This exponent appears as a nonlinear eigenvalue in the self similar solution of the microscopic decay problem. This decay problem is manageable, because the inelastic merging events are much less frequent than the pure elastic interactions, so that the vortex system has enough time to reach almost statistical equilibrium, the merging process being only a small correction then, rather like the flow over the barrier in the Kramers theory for thermally activated barrier crossing.

To conclude this rapid survey, it is fair to say that non trivial ideas of statistical physics have been introduced in the study of 2D turbulence and have lead to a basic understanding of many, if not almost all the observations (experiments are not reported here, as they appear elsewhere in this dictionary). This shows also that there is probably still a long way to go before to understand 3D turbulence.

Bibliography

[1] L. Onsager, Nuovo Cimento suppl. **6**, 279 (1949).

[2] R.H. Kraichnan, R. Montgomery, Reports on progress in physics **45**, 547 (1982).

[3] E.A. Novikov, Yu.B. Sedov, Sov. Phys. JETP **48**, 440 (1978).

[4] R. Comte, " Etudes d'écoulements bidimensionnels tourbillonaires : mouvement de 3 et 4 tourbillons ponctuels et nappes tourbillonaires ", Ph.D. Thesis, June 79, Université Pierre et Marie Curie.

[5] H. Aref, Annual review of fluid mechanics **15**, 345 (1983).

[6] H. Aref, N. Pomphrey, Proc. Roy. Soc. London, Ser. A **380**, 359 (1982).

[7] B. Jancovici, Phys. Rev. Letters **46**, 386 (1981).

[8] J. Miller, P.B. Weichman and M.C. Cross, Phys. Rev. A **45**, 2328 (1992).

[9] Y. Pomeau, Nonlinearity **5**, 707 (1992).

[10] J. Miller, Phys. Rev. Letters, **65**, 2137 (1990).

[11] R. Robert, CRAS Paris série I **309**, 757 (1989); R. Robert and J.Sommeria, J. of Fluid Mechanics **229**, 291 (1992).

[12] J. Sommeria, communication at the ASI meeting.

[13] R. Robert and J. Sommeria, Phys. Rev. Letters **69**, 2776 (1992).

[14] J. McWilliams, Phys. of Fluids A2, 547 (1990); J.R. Herring and J. McWilliams, J. of Fluid Mechanics **153**, 229 (1985).

[15] G.F. Carnevale, J. McWilliams, Y. Pomeau, J.B. Weiss and W.R. Young Phys. Rev. Letters,**66**, 2735 (1991), and Phys. of fluids A4, 1314 (1992).

[16] G.K. Batchelor, Phys. of Fluids suppl. II,**12**, 233 (1969).

[17] Y. Pomeau, Communication at the conference "Turbulencia", Aguadulce, Spain (July 1993), M. G. Velarde editor.

STRUCTURE FUNCTIONS
(in 3-D turbulence)

Y. GAGNE and E. VILLERMAUX
Institut de Mécanique de Grenoble
L.E.G.I./ U.J.F./ I.N.P.G./ C.N.R.S.
BP 53X, F-38041 Grenoble cedex, France

Historically, the concept of the structure function in 3D turbulence has been introduced by Kolmogorov[1], because, as it will be shown below, it is a powerful tool to study scaling properties of the velocity field. In particular, the famous "$k^{-5/3}$" law for the energy spectrum, in the limit of infinite Reynolds number, comes from the corresponding "$r^{2/3}$" law in the physical space. In fact, the measurements of the velocity structure functions (made in the seventies) have globally confirmed the K41 theory, but have also revealed the existence of a *small scale intermittency* which is one of the main features of 3D turbulence which is not yet understood.

In this paper, we take *a priori* the point of view of the experimentalist. First, we give some definitions and properties of the structure functions compared with the correlation functions, second, we stress the specific experimental difficulties linked with the measurements of the high order velocity structure functions, and finally we expose the main experimental results on the small scale intermittency of the velocity field which are challenges for theoreticians as well.

STRUCTURE FUNCTION AND CORRELATION

In high Reynolds turbulent flows, one quite frequently encounters random fields which are not stationary in time but are only *stationary by increments*. Consider, for instance, a diffusion process of particles in an homogeneous and steady turbulent flow.

The coordinate $x(t)$ of a particle is obviously a non stationary process, however, the increment $\delta x(\tau) = x(t + \tau) - x(t)$ is stationary since for a fixed time delay τ this increment is independent of time t. In real high Reynolds turbulent flows, the velocity $u(t)$ measured at a given point is generally a stationary increment process and thus, the velocity increment $\delta u(t, \tau) = u(t + \tau) - u(t)$ depends only on the time delay τ. Hereafter, we assume without loss of generality that such a process satisfies the condition $< \delta u(\tau) > = < u(t + \tau) - u(t) > = 0$.

Following Kolmogorov[1], the second order statistical moment of the velocity increment $\delta u(\tau)$ which is defined by

$$< (\delta u(\tau))^2 > = < [u(t+\tau)-u(t)]^2 > = S_2(\tau)$$

is called the *velocity structure function*.

Of course, other statistical moments of any order p of the random process $u(t)$ can be defined by

$$< (\delta u(\tau))^p > = < [u(t+\tau)-u(t)]^p > = S_p(\tau)$$

Strictly speaking, the $S_p(\tau)$ functions are the statistical moments of order p of the increments $\delta u(\tau)$; however, these $S_p(\tau)$ functions are often called *structure functions of order p* or *high order structure functions* of the process $u(t)$.

In the same way, one can define the spatial structure function for a random process with homogeneous increments by

$$< (\delta u(r))^2 > = < [u(x+r)-u(x)]^2 > = S_2(r)$$

In fully developed turbulence, it corresponds to the important concept of *locally homogeneous random field*.

For the study of the 3D locally isotropic turbulent velocity field, particular velocity structure functions have been introduced (cf. Monin and Yaglom[2]) corresponding to the following situations:

- the *longitudinal velocity structure function* defined by

$$S_{2\parallel}(\ell) = < [u(x+\ell) - u(x)]^2 >$$

in which the velocity component u is aligned with the direction of the separation ℓ.

- the *lateral velocity structure function* defined by

$$S_{2\perp}(\ell) = < [v(x+\ell) - v(x)]^2 >$$

in which the velocity component v is normal to the direction of the separation ℓ.

Also, high order longitudinal and/or lateral velocity structure functions have been introduced and noted as $S_{p\parallel}(\ell) = < [u(x+\ell) - u(x)]^p > .$

Let us make some remarks:

- By definition, the structure function $S_2(\tau)$ is always nonnegative, even and such that $S_2(0) = 0$.

- The structure function $S_2(\tau)$ is as important for the study of non stationary processes with stationary increments as the classical correlation function $c(\tau) = < u(t-\tau)u(t) >$ for stationary fields; in the particular case of a stationary field, the structure function $S_2(\tau)$ can be expressed in terms of the correlation $C(\tau)$, and we have

$$S_2(\tau) = 2(C(0) - C(\tau))$$

In turbulent flows, the velocity correlation $C(\tau)$ generally vanishes as τ increases, then one has $S_2(\infty) = 2C(0)$; therefore, in that case, the correlation and the structure function are interchangeable:

$$C(\tau) = 1/2(S_2(\infty) - S_2(\tau))$$

- Experimentally, ensemble averages are replaced by time averages over a long sequence of the time interval τ. In the case of a stationary field, it turns out that the time convergence of $S_2(\tau)$ is more easily reached than the correlation one. This is due to the fact that the velocity increment $\delta u(\tau)$ has no spectral component periods much greater than τ. Correlation functions are sometimes estimated with the structure function even for stationary processes.

EXPERIMENTAL DATA

Fully developed turbulent flows are characterized by a wide hierarchy of scales ℓ. The relevant experimental quantity to study the velocity statistics of inertial structures are the velocity increments $\delta u(\ell)$. This difference of velocity measured at two points separated by a distance ℓ gives roughly the instantaneous local velocity u_ℓ of an "eddy" of size ℓ. By varying the separation ℓ from the integral scale ℓ_0 (the largest one) to the dissipation scale ℓ_d (of order of the Kolmogorov scale), experimentalists studied the velocity scaling properties in the inertial range of different types of flows (channel, jet, boundary layer, wake, shear layer), for different Taylor scale based Reynolds numbers ($20 \leq R_\lambda \leq 13000$) (cf. Van Atta and Chen[3]; Van Atta and Park[4]; Antonia et al.[5], Anselmet et al.[6], Praskovsky et al.[7]). To day, with the huge capacity of computers, some numericists such as Vincent and Meneguzzi[8] begin to study the structure functions of the velocity field calculated by direct numerical simulation of the Navier Stokes equations corresponding to Reynolds number of order of $R_\lambda \simeq 150$.

Measurement Techniques

By definition, the existence at a given separation ℓ of the high order velocity structure function $S_{p\parallel}(\ell)$ is mainly governed by the shape of the tails of the probability density function (hereafter noted pdf) of the velocity increment $\delta u_\parallel(\ell)$ since we have:

$$S_{p\parallel}(\ell) = \int P(\alpha)\alpha^p d\alpha$$

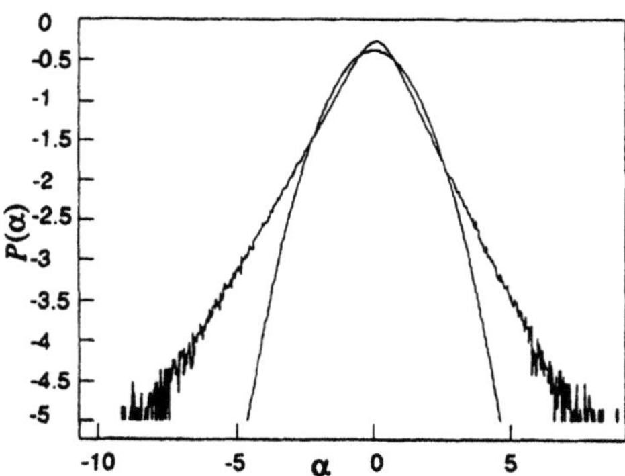

Figure 1: Pdf of the normalized velocity increment $\delta u(\ell)$ measured in an axisymmetric jet at a moderate Reynolds number $R_\lambda \simeq 850$. Here, the separation is $\ell \simeq 10\eta$ (η is the Kolmogorov scale characteristic of the dissipation range). The parabola corresponds to the Gaussian distribution.

Here α is the normalized velocity increment $\delta u(\ell)/[< (\delta u(\ell))^2 >]^{1/2}$ with $< \delta u(\ell) >= 0$ and $P(\alpha)$ is the pdf of α.

In high Reynolds turbulent flows, experimental data of the distribution of $\delta u(\ell)$ show that the tails of the pdf decrease faster than any power law suggesting that the $S_{p\parallel}(\ell)$ of any order p can be measured. This experimental fact is shown on Fig. 1 which displays an experimental pdf of $\delta u(\ell)$ measured in a classical laboratory jet at $R_\lambda \simeq 850$ with quasi-exponential tails. Note that a few experimental results such as those obtained in the atmospheric flows by Lovejoy and Schertzer[9] suggest that all the velocity structure functions of order greater than 5 or 6 diverge. Here, we will assume that all the $S_{p\parallel}(\ell)$ functions exist, which is generally accepted.

The specific difficulties associated with the measurement of the velocity structure functions have been discussed by Tennekes and Wyngaard[10]. The main problem concerns the estimation of the integrand $\alpha^p P(\alpha)$ which yields the value of the statistical moments at a given separation ℓ. Fig. 2 gives, for instance the behavior of the term $\alpha^{12} P(\alpha)$ deduced from the pdf shown on Fig. 1. The main contribu-

tion is given by the negative fluctuations of α. This is a generic feature due to the intrinsic negative asymmetry of the velocity distribution of any developed turbulent flow. On Fig. 2, it is also clear that the computation of the twelfth order moment requires the measure of the fluctuations greater than 13 times the standard deviation corresponding to very rare events (less than 10^{-6} in probability). Therefore, experimentalists need: (i) a very large dynamic range in the instrumentation (especially for the A/D converter) in order to catch the strongest events, and (ii) a very long time signal in order to achieve a good statistical convergence of these dynamical events. Such requirements imply the use of large files of sampled velocity signal, typically 10^8 data for the 14th order at moderate Reynolds number. Experiments showed that the statistical convergence, at a given scale ℓ, is worse for odd order $(S_{2p+1\parallel}(\ell))$ than for the adjacent even one $(S_{2p\parallel}(\ell))$.

On the other hand, at a fixed order p, the convergence is easiest for the large separation ℓ corresponding to the largest scale of the inertial range of the velocity field (Antonia and Van Atta[11], Antonia et al.[5]).

Experimental results

Here we will focus only on the velocity field. Experimental (physical and numerical) data gathered during the thirty last years indicate that the velocity structure functions $S_{p\parallel}(\ell)$ have scaling properties for the separation ℓ lying in the inertial range:

$$< (\delta u(\ell))^p >= S_{p\parallel}(\ell) \sim \ell^{\zeta_p}$$

Fig. 3 shows the typical power law behavior of the structure function $S_{p\parallel}(\ell)$ corresponding to the case of the Figs. 1 and 2.

Van Atta and Chen[3] and Van Atta and Park[4] were the first to give the experimental behavior of the scaling exponents ζ_p. Fig. 3 emphasizes the importance of an objective criterion for determining the inertial range. A relatively small shift in the inertial zone can significantly affect

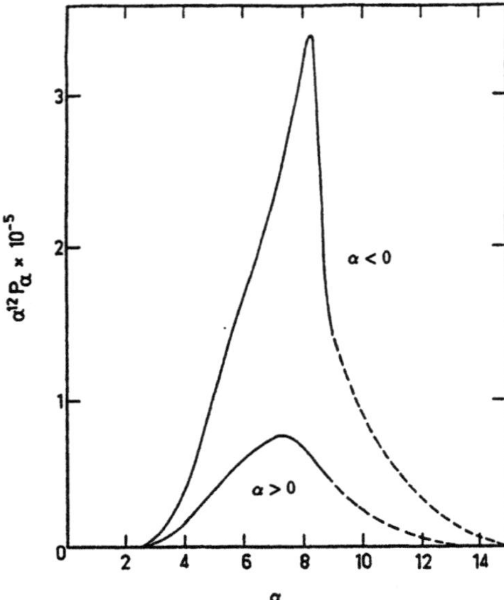

Figure 2: Shape of the 12th order integrand $\alpha^{12} P(\alpha)$ measured in a jet for $R_\lambda \simeq 540$. Here, the separation is $\ell \simeq \lambda$, (λ is the Taylor scale characteristic of the inertial range). Broken lines are extrapolations of solid lines beyond the experimental range of α.

the slopes. Here, the limits of the inertial range were determined by the third order velocity structure function $S_{3\parallel}(\ell)$ which is free of any intermittency assumption, and behaves as

$$S_{3\parallel}(\ell) = -4/5 < \varepsilon > \ell$$

In Fig. 3, it appears that this definition is very stringent, since most of the structure functions for $p \leq 12$ seem to display a power law behavior on a more extended range. On the other hand, we see that all the structure functions are oscillating along the inertial scales, especially for high order. This feature is always observed and does not seem connected to a poor statistical convergence. To day, these oscillations are not yet completely understood.

Despite all these uncertainties, it is possible to estimate the average value of the slopes of the $S_{p\parallel}(\ell)$ functions leading to the experimental values of the statistical scaling exponents ζ_p. In particular, it is better

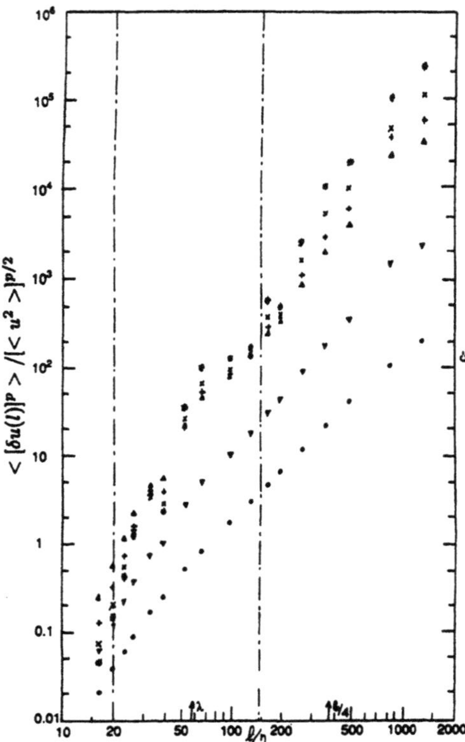

Figure 3: Inertial range behavior of structure functions $< [\delta u(\ell)]^p >$ with ℓ/η: $\circ$ $p = 8$; ∇ 10; $\triangle$ 12; $+$ 14; $\times$ 16; $*$ 18. (The 14th order is divided by 50, 16th by 10 and 18th by 10^3). The vertical chain dotted lines indicate the inertial range limits defined by the "Kolmogorov law" of the third order structure function.

to use the slopes of the high order structure functions in the real space than to measure the ζ_p exponents with the slopes of the inertial range of the high order spectra. In fact, the pure power law zones of spectra are more affected by the edges (the integral scale and the viscous cut-off) than those of the corresponding physical space structure functions (Nelkin[12]). Fig. 4 gives the behavior of these exponents as a function of the order p for different turbulent flows, and clearly shows that the ζ_p exponents have a non linear dependence on the order p. This experimental result is fundamental in fully developed turbulence. First, it proves that the concept of *global scaling invariance* assumed by Kolmogorov 1941 in the limit of infinite Reynolds number is not completely verified. Even at very high Reynolds number, experimental data suggest that only a *local scaling invariance* (like in the multifractal model of Parisi and Frisch[13]) can be assumed. Second, the non linear behavior shows that the shape of the pdf of the velocity increment $\delta u(\ell)$ is not self-similar when the scale ℓ varies from the integral scale ℓ_0 to the dissipation scale ℓ_d. This fact is one of the main problems in order to take into an account the small scale intermittency in fully developed turbulence. Third, one can observe that the values of the ζ_p exponents increase monotonically with the order p, but if one distinguishes the odd values ζ_{2p+1} and the even ones ζ_{2p}, one sees that these two sets of exponents follow two distinct functions of the order p. This point has been experimentally emphasized (Van der Water[14]) and has been recently interpreted in terms of the multispiral model (Vassilicos[15]). For the comparison between statistical models of small scale intermittency and experimental data, the reader can see other entries of this book.

Recently, Benzi *et al.*[16] introduced the concept of extended self-similarity based on the self (or relative) scaling of two distinct structures functions. From the scaling property of two structure functions of order p and q (namely, $S_{p\|}(\ell) \sim \ell^{\zeta_p}$ and $S_{q\|}(\ell) \sim \ell^{\zeta_q}$ for the inertial scale ℓ), one can study the following self power law behavior

$$S_{p\|}(\ell) \sim [S_{q\|}(\ell)]^{\beta(p,q)}$$

with $\beta(p,q) = \frac{\zeta_p}{\zeta_q}$. The main result is that this self scaling is experimentally verified not only for the inertial scale ℓ, but also for the dissipation scale ℓ up to the Kolmogorov scale η. This means there is some self (or relative) scaling at a given scale ℓ of the dissipation range, even though there is no direct (or absolute) scaling in this range.

Currently, some interesting questions on velocity structure functions arise. On the

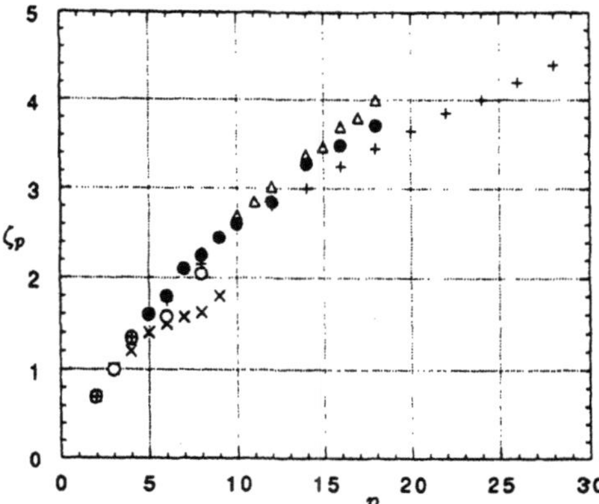

Figure 4 : Dependence of scaling exponents ζ_p on the order p. Physical experiments: $\triangle$ grid turbulence $R_\lambda \simeq 260$; o jet $R_\lambda \simeq 500$; • jet $R_\lambda \simeq 850$; × atmospheric boundary layer $R_\lambda \simeq 3000$. Numerical experiment: + isotropic turbulence $R_\lambda \simeq 150$.

one hand, there is a lack of physical experimental data on the lateral velocity structure functions $S_{p\perp}(\ell)$, and it will be very useful to confirm the few numerical experimental results (obtained at low Reynolds number) which suggest that at a given order p, the scaling exponents $\zeta_\|^p$ are different from the $\zeta_\perp^p$ ones (Kerr[17], Brachet[18]). On the other hand, an important issue is to check if the scaling properties of the velocity structure functions can be generalized to all the scales of the turbulent field (inertial and dissipation ranges), with some kind of extended absolute scaling by using a universal function f like it has been recently proposed (Benzi *et al.*[16], Villermaux and Gagne[19]):

$$S_{p\|}(\ell) \sim [\ell f(\ell/\eta)]^{\zeta_p}$$

Despite their own interest, passive scalar structure functions have been less experimentally investigated but are, anyway, out of the scope of this paper.

Bibliography

[1] Kolmogorov, A.N. 1941 *Dokl. Akad. Nauk SSSR*, **30**, 299-305.

[2] Monin, A.S. & Yaglom, A.M. 1975 in *Statistical Fluid Mechanics*, vol. 2, MIT Press.

[3] Van Atta, C.W. & Chen, W.Y. 1970 *J. Fluid Mech.*, **44**, 145-159.

[4] Van Atta, C.W. & Park, J. 1972 in *Statistical Models and Turbulence*, Lecture Notes in Physics, vol. 12, 402-426. Springer-Verlag.

[5] Antonia, R.A., Satyaprakash, B.R. & Chambers, A.J. 1982 *Phys. Fluids*, **25**, 29-37.

[6] Anselmet, F., Gagne, Y., Hopfinger, E.J. & Antonia, R.A. 1984 *J. Fluid Mech.*, **140**, 63-89.

[7] Praskovsky, A.A. 1993 in *Annual Research Briefs*, CTR-NASA Stanford, 269-276.

[8] Vincent, A. & Meneguzzi, M. 1991 *J. Fluid Mech.*, **225**, 1-20.

[9] Schertzer, D. & Lovejoy, S. 1984 in *Turbulence and Chaotic Phenomena* , 505-508. Elsevier-North Holland.

[10] Tennekes, H. & Wyngaard, J.C. 1972 *J. Fluid Mech.*, **55**, 93-103.

[11] Antonia, R.A. & Van Atta, C.W. 1978 *Phys. Fluids*, **21**, 1096-1099.

[12] Nelkin, M. 1981 *Phys. Fluids*, **24**, 556-557.

[13] Parisi, G. & Frisch, U. 1984 in *Turbulence and Predictability in Geophysical Fluid Dynamics and Climate Dynamics*, 84. North-Holland.

[14] Van der Water, W. 1993 in *Turbulence in Spatially Extended Systems*, Les Houches.

[15] Vassilicos, J.C. 1992 in *Topological Aspects of the Dynamics of Fluids and Plasmas*, 427-442. Kluwer Academic Press.

[16] Benzi, R., Ciliberto, S., Tripiccione, R., Baudet, C., Massaioli, F. & Succi, S. 1993 *Phys. Rev. E*, **48**, 29.

[17] Kerr, R.M. 1985 *J. Fluid Mech.*, **153**, 31-58.

[18] Brachet, M. 1990 *C.R.Acad.Sci. Paris*, **311**, Serie 2, 775-780.

[19] Villermaux, E. & Gagne, Y. 1993 *Phys. Rev. Lett.*, **73**, 2, 253-255 (1994)

Vorticity Filaments

Y. Couder, S. Douady and O. Cadot
Laboratoire de Physique Statistique de l'E.N.S.,
24 rue Lhomond, F-75231 Paris cedex 05, France

Since such early works as those by Townsend[1] and Kuo and Corrsin[2] it is only recently that a renewed interest for the spatial structures of turbulent flows has appeared. One of the reasons lies in the emergence of new possibilities of investigation; in particular the increase in power of the recent computers now permits to perform direct numerical simulations with a resolution compatible with the multiscale characteristic of the turbulence.

The two main ways of investigating turbulent flows, experiments and numerical simulations, have severe but different limitations. In experiments, where very high Reynolds numbers can be reached, the measurements are usually limited to those done in one point of the flow and few visualization techniques are available. In contrast, in simulations, measurements can be done everywhere and visualization of the various fields is easier but the Reynolds numbers are rather low and a continuous observation of the flow during long periods is not possible. For this reason experiments and simulations both brought about incomplete information.

The comparison of their results however led to agreement for the definition of well defined dynamical structures. Though their characterization is not completed yet and their exact role in the statistics of turbulence is still somewhat controversial, we will summarize some recent results obtained about two of them: the layer-shaped stretched shear regions and the vorticity filaments.

TOOLS FOR THE OBSERVATION OF THE VORTICITY FILAMENTS AND OF THE LAYER SHAPED SHEAR REGIONS

In numerical simulation[3-13], as it is possible to visualize directly various fields, a characterization of the inhomogeneity of turbulence in space and time is often obtained by studying the repartition of two dynamically important quantities: the vorticity modulus ω^2:

$$\omega^2 = 1/2 \sum_{ij} (\partial_i v_j - \partial_j v_i)^2 \qquad (1)$$

and σ^2 the square of the rate of strain tensor:

$$\sigma^2 = 1/2 \sum_{ij} (\partial_i v_j + \partial_j v_i)^2 \qquad (2)$$

this latter quantity being directly linked with the energy dissipation $\epsilon = \rho \nu \sigma^2$, where ρ is the fluid density, and ν its kinematic viscosity

Experimentally, it is not, at present, possible to visualize directly the vorticity or the dissipation. However Douady et al.[14, 15] introduced recently a new technique which permits the direct observation of the regions of minimum pressure. This possibility is interesting in view of the relation of the pressure field p with the vorticity and the dissipation. By taking the divergence of the Navier-Stokes equations, in a constant density, incompressible fluid, we find:

$$2\Delta p/\rho + \sigma^2 - \omega^2 = 0 \qquad (3)$$

Turbulence: A Tentative Dictionary
Edited by P. Tabeling and O. Cardoso, Plenum Press, New York, 1995

This equation is analogous to Poisson's law in electrostatics, the pressure corresponding to the potential resulting from negative and positive charges distributed in proportion to the square of the vorticity and to the energy dissipation, respectively. The vorticity concentrations thus act as sources of low pressure and wherever vorticity is separated from dissipation it will create a depression. The basic idea of the experiment is that it is possible to single out and visualize the regions of large vorticity and small dissipation because buoyant particles such as little bubbles will migrate and collect into the regions of minimum pressure. This technique can be used in most types of non thermal turbulent flows.

DESCRIPTION OF THE VORTICITY FILAMENTS

Siggia[3] was the first to show in a 64^3 simulation that in a turbulent flow the regions in which vorticity had the largest modulus were in the shape of elongated filaments (or tubes). Simulations by Brachet[4, 5], while they confirmed a repartition of the vorticity in filaments, showed that the region of large energy dissipation were rather distributed on sheets. Both these results have now been confirmed and made more precise in several simulations of high resolutions due to She et al.[6, 7], Vincent et al.[8], Kida et al.[9], Tanaka et al.[10], Metais et al.[11], Jimenez et al.[10] and Pumir[13]. The variety of forcing schemes used in the different cases show that the filaments are generated by the turbulent flow whatever its type. In experiments[14, 15] using the visualization with microbubbles, the intermittent formation of regions of minimum pressure with a filamentary shape was obtained at Reynolds numbers up to 10^6. At such high values of the Reynolds number a hierarchy of filaments of different lengths was obtained, some of them being as long as the injection length scale. The direct relation of the small pressure filaments with the vorticity filaments was first shown in simulations by Brachet et al.[4, 5] and Metais et

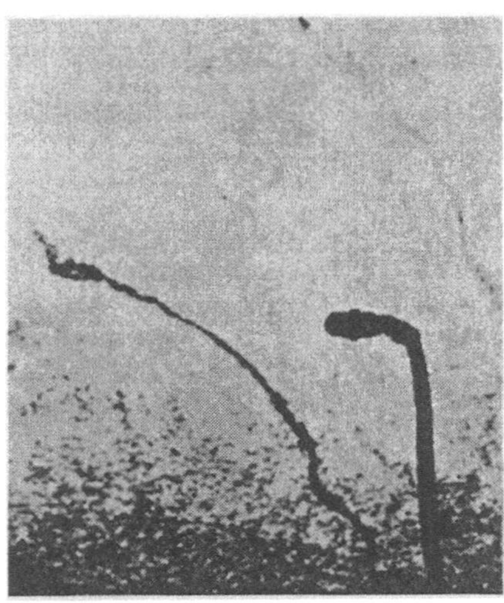

Figure 1: A filament with a length of the order of L seen from its side. The black form to the right of the picture is the pressure probe. Shorter and weaker filaments can also be observed.

al.[11]. It was more quantitatively defined in a work by Tanaka and Kida[10]. In a simulation of a turbulent flow they investigated the correlation between the distribution functions of vorticity and dissipation. They showed that there was an asymmetry of these diagrams. Large vorticity appeared associated with either weak or large dissipation, while large dissipation was always associated with large vorticity. By imposing threshold values they demonstrated that the regions of large ω^2 and weak σ^2 were in the shape of filaments. These are the regions of low pressure observed in the experiment. In contrast the regions with both large σ^2 and ω^2 are flat in the shape of pancakes. This asymmetry in the vorticity and dissipation distributions is reflected in the histograms of the pressure[5, 11, 16, 17], a point to be discussed below.

The question of the sizes of the filaments is not entirely decided yet. The simulation by Jimenez et al.[12] suggests that the filaments can have simultaneously a length of the order of the injection scale and a core diameter as small as the Kolmogorov dissipation scale. However in their simulations the long filaments are only obtained by reconnecting small filaments together. Novikov[18] on theoretical grounds proposed that there existed another new length scale in turbulence $LRe^{-3/10}$ that could characterize the filaments. In the experiment, as stated above, filaments with a hierarchy of lengths are observed (see Fig. 1). Cadot *et al.*[17] suggest that there is a corresponding hierarchy in their transverse dimension. In their model the diameter of the longest of the filaments is of the order of the Taylor microscale $LRe^{-1/2}$ and only the shorter filaments have a core size down to the dissipative scale. Their argument will be given in the paragraph about the formation of the filaments.

DYNAMICS OF THE FILAMENTS

Some information exists about the formation and disappearance of vorticity filaments, again coming partly from the experiments and partly from the simulations.

Formation

In the experiment the filaments are observed to form abruptly in a time of the order of one large eddy turn-over time and to be rather straight at the time of their formation. Fig. 1 shows their aspect as seen from the side. Filaments of various lengths and widths are observed, the largest being as long as the injection length scale L. As the bubbles also serve as a passive scalar for large scale motion it is possible to inspect the regions where the filaments form. The filaments appear abruptly: no noticeable vortex preexists when and where they show up. They rather correspond to very early stages of the roll up of shear layers which are also stretched along the direction of vorticity. The formation of such layers itself has been observed in the numerical simulation of the Eulerian evolution of a Taylor-Green vortex as well as that of other flows[19].

These simulations show the spontaneous formation of flat pancake-shaped regions in which the shear becomes spontaneously increasingly concentrated. This observation is to be related with a previous analysis of the eigenvalues of the strain tensor. Ashurst *et al.*[20] and Kerr[21] have demonstrated that the statistically dominant situation in a turbulent flow corresponds to a situation where two of the eigenvalues have a large modulus and opposite signs while the third one is only weakly positive. This situation is precisely that of a shear layer (with the two large eigenvectors at 45° from the plane of the layer) which is stretched in a direction parallel to the vorticity. In the eulerian simulations however, it is not possible to observe the long time behavior of these layers because their thickness becomes smaller than the resolution of the computation. The experimental observation[14, 15] shows that in the viscous situation these layers roll up, a process which is responsible for the formation of the filaments. There are two possible processes for this roll up: either the simple Kelvin-Helmholtz instability, or a mechanism of collapse first investigated by Lin and Corcos[22] and J. Neu[23]. These two latter works were done in the context of the secondary instability of the main eddies of a shear layer where they show that in the presence of vorticity stretching there is a collapse of the vorticity layers into narrow filament-like vortices. Moffat *et al.*[24] have recently generalized this investigation of the stretching to all the non axisymmetric strain fields.

The argument put forward by Cadot *et al.*[17] is the following. They first consider that the shear layers of length l_f (along the direction of vorticity) are created at the periphery of energy carrying eddies having a diameter of the order l_f. In this situation it is possible to obtain a scaling argument for the thickness of these layers. A scale l_f of the turbulent motion can be written as a function of the injection length scale L as $l_f \sim LRe^{-3\alpha/4}$ where α is a parameter which ranges from $\alpha = 0$ at the injection scale L to $\alpha = 1$ at the Kolmogorov dissipation scale η. If U is the

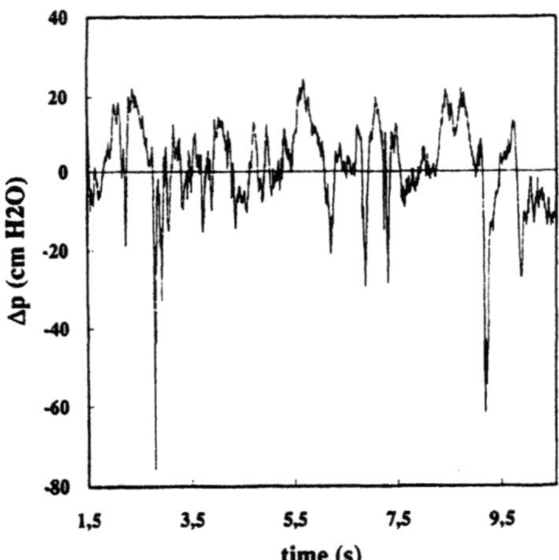

Figure 2: A time series of the pressure in one point. Two large depressions corresponding to filaments passing on the probe have been recorded.

injection scale velocity, the typical velocity at scale l_f is $v \sim U(l/L)^{1/3} \sim U Re^{-\alpha/4}$. The stretching of a vorticity layer is limited by viscosity, leading to a limiting thickness $(\nu/\gamma)^{1/2}$ where $\gamma = v/l_f$ is the stretching which can be generated at that scale. The vorticity layers created by an eddy of scale l_f will thus be of thickness $d_f \sim L Re^{-1/2-\alpha/4} \sim l_f Re^{-1/2+\alpha/2}$. The filaments result from the roll up of these layers. However in this model only a limited width (of order d_f) of the layer actually rolls up before the filament becomes unstable to vortex breakdown. The diameter of the filaments thus scales as d_f. The largest ones (due to a stretching at the largest scale L of the flow) have a transverse size $L Re^{-1/2}$, i.e. the Taylor scale λ of the flow. In contrast, near the dissipation length scale η, the diameter of the filaments becomes of the same order of magnitude as their length.

Disappearance of the filaments

The disappearance of the filaments can be difficult to observe in the simulation because it involves a change of the vorticity level. In several cases there appears to be a destructive interaction between different structures. In contrast the experiment suggests that there is an intrinsic instability of the filaments. The largest amongst them only exist during about one large eddy turnover time, after which helical distortions are observed (visible on Fig. 2). These distortions lead to the disintegration of the filaments. In the case where the filament is perpendicular to the cell wall, it is slightly stabilized so that its lifetime is somewhat longer. In this case, one can observe, after the filament's disappearance, that large long-lived eddies remain in their place. These eddies have lifetimes of the order of ten times the turnover time. This bursting process appears similar to the spiral vortex breakdown[25] which affects the vortices behind delta wings or the swirling flow in pipes of increasing diameter. In this latter case the process can be investigated in steady and controlled conditions. An experimental characterization of the flow of the vortex before breakdown shows it has the transverse velocity profile of a Burgers vortex with a concentrated jet along its core. It was shown that the instability could be triggered either by a reduction of the axial velocity or an increase of the circulation around the vortex. For the filaments which are stretched along their axis and where the circulation Γ increases due to the roll-up of the shear layer, their instability could be due to either of these two reasons. The experimental and numerical investigation of vortex breakdown in confined geometries has demonstrated several characteristics which are particularly interesting in the context of turbulence. In particular what happens in the transitional region was observed in a recent numerical simulation of the vortex breakdown by P. Mège[26] who investigated the pressure distribution across the vortex. Before the breakdown there is a smooth profile with a strong depression on the axis due to the vorticity concentration. In the breakdown region the pressure is irregularly distributed, has a larger mean value and zones of positive pressure show up. Because of Eq. (3) this is a clear indi-

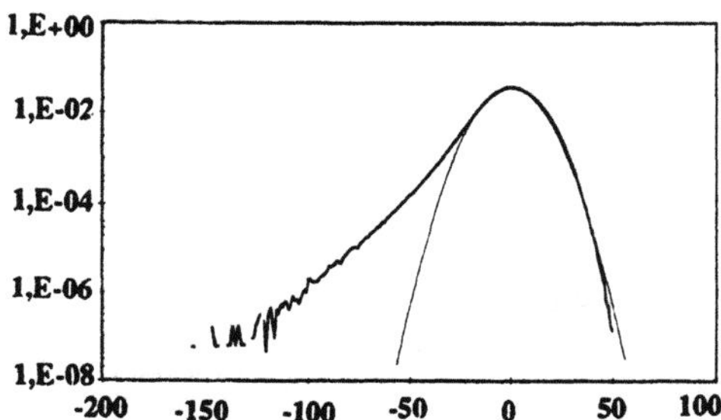

Figure 3: A histogram of the values of the pressure in one point, showing the fit by a gaussian distribution on the high pressure side and the exponential distribution of the low pressure side.

cation of the sudden formation of dissipative regions embedded in the flow. This is expected as the total energy of the mean flow upstream is larger than downstream, the difference being transferred to fluctuations which ultimately are dissipated.

One has to be cautious however in comparing these results with those in a turbulent flow because a filament embodied in a turbulent flow has more possibilities of exchange with the surrounding fluid than a single unstable vortex confined in a tube. Two features however appear comparable in the two situations. Firstly, in the turbulent flow, the instability of the filaments leads from a concentrated energetic structure to a more widely spread one (the eddy). Secondly, during this instability, the bubbles forming the filament are violently dispersed, meaning the disappearance of the region of minimum pressure and presumably the formation of dissipative regions. This can be understood in the following way. In the breakdown process the filament becomes highly contoured and different regions along its length start interacting as they approach each other. The interaction of non parallel vortices has been investigated by Boratav et al.[27] and shown to create locally strong dissipative regions in the intermediate space. The interaction of several segments of the same vortex certainly has the same effect.

Pressure distribution

Brachet[5] was the first to point out that the histograms of the pressure in all points of the flow had a strong asymmetry with a long tail only on the low pressure side. From the numerical simulations[5, 10, 11] this asymmetry in the pressure distribution appears to result from the above mentioned difference in the spatial repartition of ω^2 and σ^2. There exist regions of large ω^2 with small σ^2 (the filaments of very low pressure). But in the regions where σ^2 is large the pressure is not large because of the coexistence of large ω^2.

In the experiments the time series of the pressure measured in one point (Fauve et al.[16] and Cadot et al.[17]) exhibit the coexistence of a random looking part of moderate amplitude with large peaks exclusively oriented towards the low pressure (Fig. 2). Experimentally it is possible to record the pressure in one point while visualizing the flow with the bubble technique. Cadot et al.[17] observe direct correspondence between the measured low pressure peaks and the observation of a filament passing by the point of measurement. The corresponding histograms (Fig. 3) were investigated in detail. Fauve et al.[16] suggest an exponential tail for the low pressure regions and a power law distribution $N(p) \propto (p_+ - p)^4$ for the high pressure side. Cadot et al.[17] confirm the exponential dependence for the low pressure side but find a gaussian distribution for the high pressure side.

A remarkable feature is that comparison of histograms obtained in different experimental situations show[16, 17] that they scale as V^2 the square of the injection velocity. Experiments done with fluids of various viscosities[17] and in different setups show the histograms to be only very weakly dependent over the viscosity and the injection length scale L (provided the flow is turbulent). From this result it is possible to argue that this scaling means that the large scale circulation is not preserved during the roll up as in the collapse investigated by Neu[23]. In the model of Cadot *et al.*[17] the filament is formed by the roll up of only a part of the stretched shear layer. The resulting vorticity is the same inside the filaments of all lengths: $\omega \sim (U/L)Re^{1/2}$ but the depression is largest in the longest filaments (of diameter λ) where it scales as $\Delta p \sim U^2$ in agreement with the experiment.

Velocity measurements

The most usual measurement done in turbulent flow is that of one component of the velocity in one point. From such measurements it is possible, as was first done by Grant *et al.*[27] using the Taylor hypothesis concerning the advection of the spatial structures past the point of measurement and a hypothesis of isotropy of the turbulence, to deduce an energy spectrum in good agreement with Kolmogorov's prediction. More recently such velocity time series were reconsidered and shown[28] to exhibit intermittent bursts of velocity which contradicted the hypothesis of homogeneity of the turbulence. These bursts were ascribed to the advection, near the point of measurement, of coherent structures, but their observation by themselves gave very little information about them. Similarly, measurements of the spatial velocity gradients showed their distribution to be non Gaussian with strong exponential tails for the large values of the gradient, a result first obtained by Van Atta *et al.*[29]. These tails were also thought to correspond to the existence of coherent structures. More recently, Tsinober *et al.*[31] showed that when all the velocity components and velocity gradients are considered, the intermittent bursts in the time series of kinetic energy, local vorticity, or dissipation are much stronger and better defined than in the time series of the velocity. The previously discussed results suggest that the most intense structures in the turbulent flow are the sheet-shaped shear regions and the filaments. It is most likely that they

are responsible for the strongest bursts of these various time series.

DISCUSSION : THE ROLE OF THE VORTEX FILAMENTS IN A TURBULENT FLOW

The importance of the role that different schools of thinking ascribe to vorticity filaments is often reflected in the very name they use for them which ranges from "worms"[12] to "sinews"[24]. We suggest that the term filament is the most adequate for the description of these structures. However it carries the possibility of confusion with the weak vorticity structures observed in 2D turbulence and responsible for the dissipation of enstrophy. These latter objects are stripes or ribbons of weak vorticity. It must be pointed out that they are completely different objects since in 2D the vorticity is perpendicular to the direction of the stripe. We thus think for that clarity the term filament should be used for the above described 3D structures, while ribbons or stripes be used for the 2D case.

It is clear that the vortex filaments occupy a very limited fraction of the turbulent volume so that any instantaneous global measurement will not be very sensitive to their presence or absence. Jimenez *et al.*[12] have investigated the effect of a truncation aimed at either eliminating selectively the filaments from the flow, or eliminating all the background and only retaining the filaments. They showed the effect of the elimination of the filaments on the energy spectrum to be negligible. This absence of direct effect does not necessarily imply that they have no role. Douady *et al.*[14, 15] suggest that their role is important in the dynamical structuration of the flow: their bursting results in the formation of both large structures and dissipative regions. Support to this view could come from the possibility of giving an interpretation to the drag reducing role of the polymer. It is well known that the presence of very diluted polymers reduces the power dissipation in a turbulent flow by up to 30%. Bonn *et al.*[32] have observed that the presence of polymers appear to inhibit the formation of the filaments and thus change the global structure of the flow. In any case much more work is needed before reaching a complete understanding of the dynamics of these structures.

Bibliography

[1] A.A. Townsend, Proc. Roy. Soc. Lond. **A 208**, 534-542 (1951).

[2] A.Y.-S. Kuo and S. Corrsin, J. Fluid Mech. **50**, 285-320 (1971).

[3] E.D. Siggia, J. Fluid Mech., **107**, 375, (1981).

[4] M.E. Brachet, C.R. Acad. Sci. Paris, **311**, 775 (1990).

[5] M.E. Brachet, Fluid Dynamics Research **8**, 1-8 (1991).

[6] Z.S. She, E. Jackson, S.A. Orszag, Nature **344**, 6263, 226 (1990).

[7] Z.S. She, E. Jackson and S.A. Orszag, Proc. R. Soc. Lond. **434**, 101 (1991).

[8] A. Vincent and M. Meneguzzi, J. Fluid Mech. **225**, 1 (1991).

[9] S. Kida and K. Ohkitani, Phys of Fluids A **4**, 1018 (1992).

[10] M. Tanaka and S. Kida, Phys of Fluids A **5**, 2079 (1993).

[11] O. Metais and M. Lesieur, J. Fluid Mech. **239**, 157-194 (1992).

[12] J. Jimenez, A.A. Wray, P.G. Saffman and R.S. Rogallo, J. Fluid Mech. **255**, 65-90 (1993).

[13] A. Pumir, A numerical study of pressure fluctuations in three-dimensional incompressible homogeneous isotropic turbulence. Preprint (1993).

[14] S. Douady, Y. Couder and M.E. Brachet, Phys. Rev. Lett. **67**, 983, (1991).

[15] S. Douady, Y. Couder, in "Turbulence in Extended systems" p.3-17 R. Benzi, C. Basdevant and S. Ciliberto Editors, Nova Science Publishers (1993).

[16] S. Fauve, C. Laroche and B. Castaing, J. Phys. II France **3**, 271-278 (1993).

[17] O. Cadot, S. Douady and Y. Couder, Characterization of the low pressure filaments in 3 D turbulence, Preprint 1994.

[18] E.A. Novikov, Phys. Rev. Lett. **71**, 2718-2720 (1993).

[19] M.E. Brachet, P.L. Sulem, M. Meneguzzi, H. Politano, A. Vincent, J. Fluid Mech. **194**, 333-349 (1994).

[20] W.T. Ashurst, A.R. Kerstein, R.M. Kerr and C.H. Gibson , Phys. Fluids **30**, 3243 (1987) and W.T. Ashurst, J.Y. Chen, and M.M. Rogers, Phys. Fluids **30**, 3293 (1987).

[21] R.M. Kerr, Phys. Rev. Lett. **59**, 783 (1987).

[22] S.J. Lin and G.M. Corcos, J. Fluid Mech. **141**, 139 (1984).

[23] J.C. Neu, J. Fluid Mech. **143**, 253 (1984).

[24] H.K. Moffat, S. Kida and K. Okhitani, J. Fluid Mech. **259**, 241-264 (1993).

[25] S. Leibovich, Ann. Rev. Fluid Mech. **10**, 221 (1978).

[26] P. Mège, T.H. Le and Y. Morchoisne, in "Turbulence and coherent structures" edited by O. Métais and M. Lesieur. Kluwer Academic press (1991).

[27] O.N. Boratav, R.B. Pelz and N.J. Zabusky, Phys. Fluids A **4**, 581 (1992).

[28] H.L. Grant, R.W. Stewart and A. Moilliet, J. Fluid Mech. **12**, 241 (1962).

[29] F. Anselmet, Y. Gagne, E.J. Hopfinger and R. Antonia, J. Fluid Mech. **140**, 63-89 (1984).

[30] C.W. Van Atta and W.Y. Chen, J. Fluid Mech. **44**, 145-159 (1970).

[31] A. Tsinober, E. Kit and T. Dracos, J. Fluid Mech. **242**, 169-192 (1992).

[32] D. Bonn, Y. Couder, P.H. van Dam and S. Douady, Phys. Rev. **47**, R28 (1993).

WAVELET ANALYSIS
(of single-point turbulence data)

A. ARNEODO, E. BACRY† and J.F. MUZY

Centre de Recherche Paul Pascal,

Avenue Schweitzer, F-33600 Pessac, France

†also at Université de Paris VII, UFR de Mathématiques,

Tour 45-55, 2 Place Jussieu, F-75251 Paris cedex 05, France

The central problem of three-dimensional fully developed turbulence is the energy cascading process. It has resisted all attempts at a full understanding or mathematical formulation. The main reasons for this failure are related to the large hierarchy of scales involved, the highly nonlinear character of the Navier-Stokes equations and the spatial intermittency of the dynamical active regions[1, 2]. In this context, statistical and scaling properties have been the basic concepts used to characterize turbulent flows. One of the striking signatures of the so-called intermittency phenomenon is the non-gaussian statistics at small scales. The energy transfer towards small scales is related to the non-zero skewness of the probability distribution function (PDF) of the velocity increment and the large flatness of the PDF (kurtosis) corresponds to the presence of strong bursts in the energy dissipation. This fine-scale intermittency is responsible for some departure to the classical theory of Kolmogorov[3] (K41) which neglects the presence of fluctuations in the energy transfer. Mandelbrot[4] was the first one to advocate the use of fractals in turbulence. Some of his early multiplicative cascade models contained all the ingredients of the multifractal formalism[5] that has recently proved particularly fruitful in the characterization of singular objects arising in a variety of physical situations. During the past few years, considerable effort has been devoted to the multifractal analysis of high Reynolds number

turbulence[2]. But the problem of comparing the predictions of multifractal cascade models with experimental data comes from the fact that three-dimensional processing of turbulent flows is at the moment feasible only for numerical simulations which are unfortunately limited in Reynolds numbers to regimes where the scaling just begins to manifest itself. Present experimental techniques have access to the two-dimensional structure of passive scalars[6] and only to the one-dimensional cuts of the velocity field[7, 8]. Here, we are only interested in the statistical analysis of single-point data based on hot-wire techniques in the presence of a mean flow (wind tunnels, jets, etc...).

MULTIFRACTAL ANALYSIS OF TURBULENCE DATA

There are mainly two approaches to multifractals in turbulence. The first one is based on the determination of the generalized fractal dimensions D_q and the $f(\alpha)$ singularity spectrum of the energy dissipation[4, 7]. The second one consists in directly investigating the distribution of singularities of the velocity field itself via the scaling behavior of the structure functions[8, 9].

Multifractal description of the energy dissipation

Following a tradition going back to Kolmogorov[3], Mandelbrot[4] originally proposed to investigate the fluctuations of the local energy dissipation. But the data obtained from point-probe measurements give only access to the stream (longitudinal) velocity component $v(t)$. Using Taylor's frozen flow hypothesis and further assuming that the gradient of only one velocity component in one direction is representative of the dissipation, the dissipation ϵ is generally defined as[7]

$$\epsilon(x) = \left(\frac{dv(x)}{dx}\right)^2. \qquad (1)$$

The scaling behavior of ϵ (considered as a measure distributed over the real axis) has been studied over domains of size pertaining to the inertial range: $l_d << l << l_0$, where l_d is the Kolmogorov dissipation scale and l_0 the integral scale. Classical techniques (e.g. box-counting algorithms) issued from the multifractal theory have been used to extract the generalized fractal dimensions D_q and the $f(\alpha)$ singularity spectrum. The consistency of the measured spectra brings conspicuous evidence for the multifractal nature of dissipation field[7]: D_q is a monotonic decreasing function of q while $f(\alpha)$ turns out to be a single humped function. The detailed comparison performed by Meneveau and Sreenivasan[7] of their experimental results with the prediction of various cascade models clearly demonstrates the inadequacy of log-normal models[10, 11] and of cascade models with a single exponent[4, 12] including the so-called β-model[13]. Surprisingly, the simplest version of weighted curdling models proposed by Mandelbrot[4], namely the binomial model, turns out to account reasonably well (at least at a certain level of description) for the observed multifractal spectra[7]. Even though, owing to the degeneracy of the multifractal formalism, one cannot reasonably expect to extract the turbulent fragmentation process from the $f(\alpha)$-spectrum information only, the binomial models give undoubtedly the simplest possible mechanism that reproduces most of the experimental observations recorded so far on the turbulent energy dissipation.

Multifractal description of the velocity field: the structure function approach

An alternative description of the fine-scale intermittency phenomenon consists in investigating the singular aspect of the longitudinal velocity signal $v(x)$ itself[8]. This approach pioneered by Parisi and Frisch[9] relies on the determination of the spectrum $D(h)$ of Hölder exponents h of the velocity field from the inertial scaling properties of structure functions[1] (SF) of variable order:

$$S_p(l) = <(\delta v_l)^p> \sim l^{\zeta_p}$$
$$(p \text{ integer} > 0), \qquad (2)$$

where $\delta v_l(x) = v(x + l) - v(x)$ is the longitudinal velocity increment over a distance l. As later confirmed by Anselmet et al.[8], the exponents ζ_p were found to depend nonlinearly on p, deviating significantly from the prediction[3] $\zeta_p = p/3$ of K41 based on the assumption that, at each point of the fluid, the velocity field has the same scaling behavior $\delta v_l(x) \sim l^{1/3}$, which yields the well known $E(k) \sim k^{-5/3}$ energy spectrum. This observation, however, does not rule out the notion of scaling invariance which might in fact be relevant at a local level. The idea of Parisi and Frisch[9] was to interpret this nonlinearity as a direct consequence of the existence of spatial fluctuations in the scaling behavior of $\delta v_l(x) \sim l^{h(x)}$. For each h there is a set in $I\!R$ of Hausdorff dimension $D(h)$ for which $\delta v_l \sim l^h$. Thus, by suitably inserting this local scaling behavior into Eq. (2), one can bridge $D(h)$ and ζ_p by a Legendre transform

$$D(h) = \min_p(ph - \zeta_p + 1), \qquad (3)$$

which is, *a priori*, the counterpart of the Legendre transform which relates the $f(\alpha)$ and $\tau(q) = (q-1)D_q$ spectra of multifractal measures[5]. It is clear from Eq. (3), that the nonlinear dependence of ζ_p on p is equivalent to the assumption that there is more than a single scaling exponent h.

The Kolmogorov's refined similarity hypothesis

Even though there is, *a priori*, no obvious relation between the dimension $D(h)$ and the $f(\alpha)$ singularity spectrum of the dissipation, the experimental results on both the velocity structure functions[8] and the energy dissipation[7] converge to a same conclusion: a multifractal description of the intermittency of the fine structures is very appealing whereas "monofractal" cascading models seem to be definitely discarded. There has been, however, some attempt[7] to bridge the two above multifractal approaches by assuming that the local dissipation $\epsilon_l(x)$ averaged over a domain of size l and the increment $\delta v_l(x)$ are related locally by the following form[11] (in the limit $l \to 0^+$):

$$\epsilon_l(x) =^s \frac{(\delta v_l(x))^3}{l}, \qquad (4)$$

where the symbol $=^s$ means that the two quantities have the same scaling laws, so that the scaling exponents are the same for the corresponding moments of arbitrary order. Eq. (4) is essentially the global scaling assumption of K41 reformulated in terms of fluctuating local quantities. The pertinence of this local scaling relation turns out to be very much debated in recent numerical[14] and experimental[15] studies. Taking this relation for granted, one gets the following connection between the two multifractal approaches:

$$h \;=\; \alpha/3,$$

$$D(h) \;=\; f(\alpha), \qquad (5)$$

$$\zeta_p \;=\; (\frac{p}{3} - 1)D_{p/3} + 1.$$

Previous attempts to check experimentally those relations have not revealed any inconsistency[15]. But there is far from those tentative comparisons and a quantitative experimental validation.

WAVELETS AND MULTIFRACTAL FORMALISM FOR FRACTAL SIGNALS

Beyond the technical difficulties related to point-probe measurements[16], there are fundamental limitations to the previous experimental analysis. In particular, it is clear that squared derivatives of the longitudinal velocity in Eq. (1) may not be fully representative of the 3D local dissipation[17]. The situation is even worse with the velocity structure functions approach which intrinsically fails to fully characterize the $D(h)$ singularity spectrum; actually, as shown in ref. [18], only the part of the spectrum corresponding to $0 < h < 1$ is amenable to this method which fails to identify the strongest ($h < 0$) singularities as well as singularities in the derivatives of the signal ($h > 1$). Moreover, regular behavior may introduce drastic bias in the estimate of the ζ_p's (Eq. (2)). Our purpose here is to elaborate on a novel strategy that we have recently proposed[19] as an alternative method which will allow us to determine the whole $D(h)$ singularity spectrum directly from any experimental signal. This method is based on the use of a new tool introduced in signal analysis, *the wavelet transform*[20] (WT), which has proved very powerful to characterize the scaling properties of fractal objects[21].

Wavelets and local scaling properties of fractal signals

The WT is a space-scale analysis which consists in expanding signals in terms of *wavelets* which are constructed from one single function, the analyzing wavelet g, by means of dilatations and translations[20]. The WT of a signal $s(x)$ is generally a complex valued function which is defined as:

$$T_g(a, x_0) = \frac{1}{a} \int_{-\infty}^{+\infty} s(x) g^* \left(\frac{x - x_0}{a} \right) dx,$$
$$a > 0 \quad (6)$$

where g^* is the complex conjugate of g. For a large class of functions s, no information is lost since the transformation

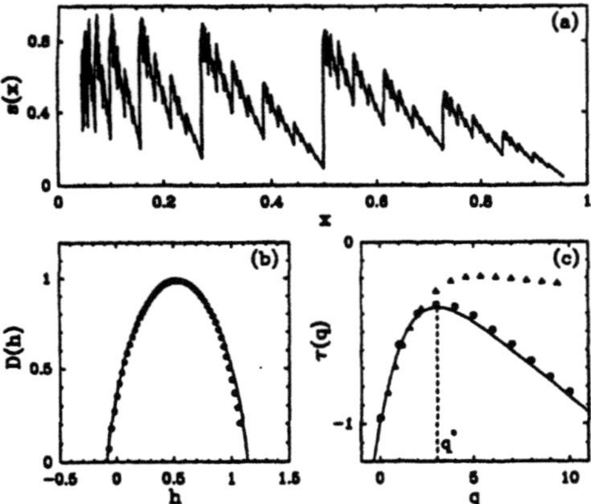

Figure 1: WTMM and SF analysis of a signal that possesses singularities of negative Hölder exponent. (a) The multifractal recursive signal. (b) $D(h)$ vs h. (c) $\tau(q)$ vs q. In (b) and (c): (•) WTMM method using the 2nd derivative of the Gaussian function as analyzing wavelet ; (▲) SF method ; (−) theoretical spectra.

is invertible provided g be well localized around $x = 0$ and have a vanishing integral. The WT can be used as a mathematical microscope[21] to analyze the local regularity of functions; g characterizes the optics while x_0 is the position analyzed and $1/a$ the magnification. In fact, if the signal $s(x)$ has, at the point x_0, a local scaling (Hölder) exponent $h(x_0)$ in the sense that, in the limit $x \to x_0$,

$$|s(x) - P_n(x_0)| \sim |x - x_0|^{h(x_0)}$$

($P_n(x_0)$ is a polynomial of degree n and $n < h(x_0) < n + 1$), then in the limit $a \to 0^+$:

$$T_g(a, x_0) \sim a^{h(x_0)} , \qquad (7)$$

provided the first $(n + 1)$ moments of g be zero[22]. Thus one can extract, from a log-log plot of the WT amplitude versus the scale a, the local scaling exponent $h(x_0)$ for any arbitrary point x_0. The situation is somewhat more intricate when investigating fractal signals. The characteristic feature of these singular signals is the existence of a hierarchical distribution of singularities. Locally, the Hölder exponent $h(x_0)$ is then governed by the singularities which accumulate at x_0. This

results in unavoidable oscillations around the expected power-law behavior of the WT amplitude[21]. The exact determination of h from log-log plots on a finite range of scales is therefore somewhat uncertain[19]. Moreover, there exist fundamental limitations (which are not intrinsic to the WT technique) to the measurement of Hölder exponents from local scaling behavior in a finite range a scales. Therefore the determination of statistical quantities like the singularity spectrum $D(h)$ requires a method which is more feasible and more appropriate than a systematic investigation of the WT local scaling behavior as experienced in refs. [23] and [24].

Determination of the singularity spectrum of fractal signals using wavelets

The classical multifractal description[5] in terms of thermodynamic quantities supposes, more or less explicitly, the existence of an underlying multiplicative process. Previous applications of the wavelet transform microscope to multifractal measures lying on Cantor sets have demonstrated its fascinating ability to reveal the construction process (renormalization operation) of recursive singular measures[21]. As proved by Mallat and Hwang[25], the WT local maxima detect all the singularities of a large class of signals. Thus the representation in the (x, a) half-plane of the local maxima of $|T_g|$ is likely to contain all the information about the hierarchical distribution of singularities of the signal. As illustrated in Fig. 2c, the WT modulus maxima are disposed on connected curves called maxima lines. Let us define $\mathcal{L}(a_0)$ the set of all the maxima lines that exist for $a \leq a_0$. Our method[18, 19] to compute the singularity spectrum of a fractal signal consists in taking advantage of the space-scale partitioning given by the maxima line skeleton to define a partition function which scales, in the limit $a \to 0^+$, in the following way:

$$\mathcal{Z}(a, q) = \sum_{l \in \mathcal{L}(a)} \left(\sup_{(x,a') \in l} |T_g(a', x)| \right)^q ,$$
$$\sim a^{\tau(q)} , \qquad (8)$$

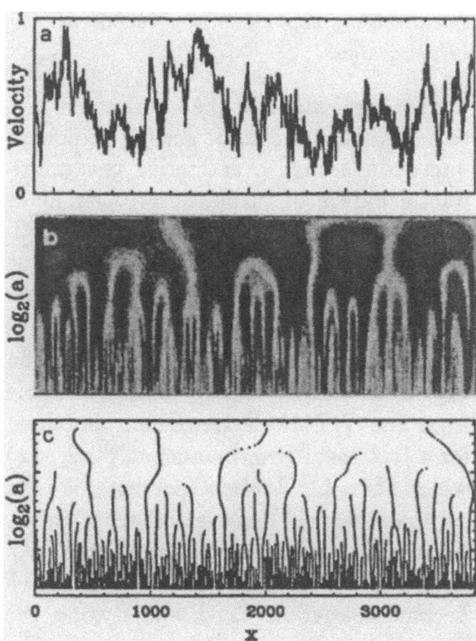

Figure 2: Continuous wavelet transform analysis of fully developed turbulence from wind tunnel data. (a) The turbulent velocity signal over about two integral length scales. (b) WT of the turbulent signal; the amplitude is coded, independently at each scale a, from white ($T_g(a,x) \leq 0$) to black ($T_g(a,x) > 0$). (c) WTMM skeleton. The analyzing wavelet is the second derivative of the Gaussian function. The large scales are at the top.

where $q \in I\!R$. By using the behavior of the wavelet coefficient along the maxima lines ($|T_g(a,x_0)| \sim a^{h(x_0)}$), one can show, in the limit $a \to 0^+$ (steepest descend method) that the $D(h)$ singularity spectrum can be computed from the exponents $\tau(q)$ via a Legendre transform[18, 19]:

$$D(h) = \min_q(qh - \tau(q)) . \qquad (9)$$

It is tempting to relate the exponents $\tau(q)$, defined from the wavelet transform modulus maxima (WTMM) method, to the exponents ζ_p of the structure functions (Eq. (2)). A simple comparison gives immediately

$$\tau(q) = \zeta_q - 1 ; \qquad (10)$$

but as discussed in ref. [18], this relationship does not hold for every value of q. In fact, it can be shown that $\tau(q)$ is a more general spectrum than the ζ_p's, in the sense that $\tau(q)$ is the exact Legendre transform of the $D(h)$ singularity spectrum in most situations[19].

In Fig. 1, we show the results of both the WTMM and SF analysis of a deterministic multifractal recursive signal that possesses some singularities with negative Hölder exponent[18]. The analytical $D(h)$ spectrum, represented by a solid line in Fig. 1b, actually extends below the value $h = 0$, over the range $[-0.1, 1.36]$. The numerical data issued from the WTMM analysis are found in remarkable agreement with the theoretical spectrum. The comparison of the WTMM and SF methods is reported in Fig. 1c where the data for $\tau(q)$ obtained from both methods are compared to the analytical spectrum. The SF numerical data systematically depart from the theoretical curve for $q > q^*$, where q^* corresponds to the critical value where $\tau(q)$ starts to decrease, i.e., where negative Hölder exponents are dominating the behavior of the structure functions (from the properties of the Legendre transform, h corresponds to the derivative of $\tau(q)$ with respect to q). This example illustrates one of the deficiencies of the SF method, i.e., its inability to resolve negative Hölder exponents. Fig. 1 shows that the WTMM method, which involves smooth analyzing wavelets, does not possess this drawback and provides a reliable quantitative characterization of the multifractality of the considered recursive signal. We refer the reader to ref. [18], where a detailed comparison of the performances of the WTMM and SF methods is reported, with special emphasis on the intrinsic limitations of the SF approach commonly used in the context of fully developed turbulence.

WAVELET ANALYSIS APPLIED TO FULLY DEVELOPED TURBULENCE DATA

In recent years, there has been increasing interest in applying the wavelet analy-

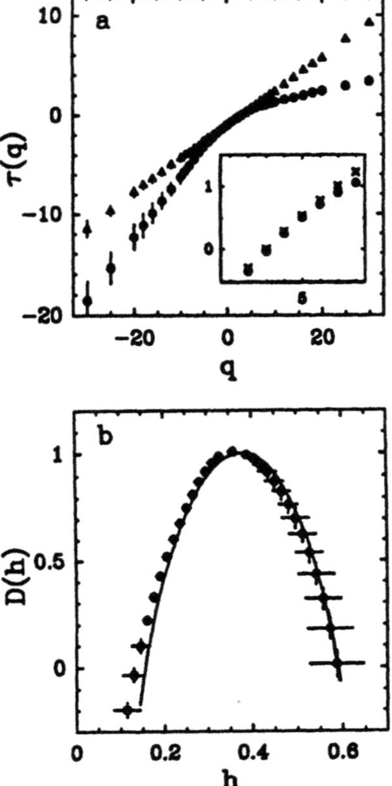

Figure 3: WTMM measurement of the singularity spectrum of fractional brownian process ($\triangle$) and turbulent velocity signal ($\bullet$). (a) $\tau(q)$ vs q. (b) $D(h)$ vs h. In (a) the symbols ($\times$) correspond to $\tau(q) = \zeta_q - 1$ obtained when computing the scaling exponents ζ_q with the SF method. In (b) the solid line corresponds to the average singularity spectrum obtained from dissipation field data[7] via the Kolmogorov scaling relation (4).

Local scaling exponents of a turbulent velocity signal

The application of the continuous WT to investigate the local scaling exponent fluctuations that characterize the multifractal nature of turbulent velocity fields at inertial range scales has been initiated in ref. [23]. Fig. 2 illustrates the wavelet transform of a sample of the velocity signal of length of about two integral scales. The WTMM skeleton in Fig. 2c is actually hardly distinguishable from the WTMM arrangement obtained for a fractional Brownian signal[4] $B_{1/3}(x)$ which has a $k^{-5/3}$ power spectrum like the turbulent signal. However, when using the additional information given by the WT amplitude (Fig. 2b), this discrimination becomes easier[19b,23]. By analyzing the behavior of $T_g(a, x_0)$ versus a along the WTMM lines, one can estimate the value of the local Hölder exponent $h(x_0)$ according to Eq. (7). Regardless of some fluctuations due to finite size effects[24], the Hölder exponent of the Brownian signal $B_{1/3}(x)$ does not depend on x: $h = H = 1/3$. In contrast, for the turbulent velocity signal, h is actually found to fluctuate in a wide range[23] between -0.3 and 0.7, thereby suggesting that the multifractal picture proposed by Parisi and Frisch[9] is appropriate. Statistically, the most frequent exponents are close to the Kolmogorov value $h = 1/3$. Let us stress the observation of negative exponents down to -0.1 and beyond, which correspond to rare but very active events. Negative exponents do not seem to have been previously reported in the literature. One possible interpretation tossed in ref. [23]b is the occasional passage nearby the probe of slender vortex filaments of the sort observed in recent experiments[27] and 3D numerical simulations[28].

Determination of the singularity spectrum of a turbulent velocity signal

In Fig. 3 are shown the results of the multifractal analysis of the Modane turbulent velocity signal performed with the

sis to turbulence data[26]. In this section, we report on the first such analysis[23] performed on single point velocity data from high Reynolds number 3D turbulence. The data where obtained by Gagne and Hopfinger in the large wind tunnel S1 of ONERA at Modane[8]. The Taylor scale based Reynolds number is $R_\lambda = 2720$ and the extent of the inertial range is almost three decades. The results reported here concern the analysis in the inertial range of about 100 integral length scales of the recorded turbulent signal (Fig. 2).

WTMM method[19]. The analysis of the Brownian signal $B_{1/3}(x)$ is shown for comparison. When plotted versus q, the scaling exponent $\tau(q)$ of the partition function (Eq. (8)) obtained for the Gaussian process, remarkably falls on a line $\tau(q) = q/3 - 1$ of slope $h = 1/3$. From the Legendre transform (Eq. (9)) one gets $D(h) = 1$. Thus, as expected theoretically, we find that the Brownian signal is everywhere singular with a unique Hölder exponent $h = 1/3$. In Fig. 3a, the $\tau(q)$ spectrum extracted from the WTMM method of the turbulent signal unambiguously deviates from a straight line. Let us note that the results obtained with the structure function method, for positive integer values of q exclusively, are in good agreement (via Eq. (10)) with the nonlinear behavior of $\tau(q)$ found with the WTMM method. The values of $h = \partial\tau(q)/\partial q$ when varying q from -30 to 30 range in the interval $[0.10, 0.62]$. The corresponding singularity spectrum $D(h)$ is shown in Fig. 3b. It displays the characteristic single-humped shape of multifractal signals. Its maximum $D(h(q = 0)) = -\tau(0) = 1.00 \pm 0.01$ does not deviate substantially from $D_F = 1$. This strongly suggests that the turbulent signal is everywhere singular[19]. The manifest part[29] of $D(h)$ (> 0) is compared to a solid curve which actually corresponds, via Eq. (5), to short-terms statistics data of dissipation fields[7] (at lower Reynolds number) assuming the validity of the Kolmogorov scaling relation (4) between the instantaneous velocity increments and the local dissipation. Despite some systematic deviations, the fact that one cannot discriminate between these two singularity spectra, within the experimental uncertainty, can be interpreted as an experimental verification of the Kolmogorov hypothesis. This observation can also be understood as an experimental confirmation of the universality of the multifractal singularity spectrum of fully developed turbulence[7]. However, it is clear that considerable further work is needed for reliable quantitative conclusions. In particular, long-term statistical analysis have to be carried out in order to capture more accurately the latent part[29] $(D(h) < 0,$ which means that the events do not occur in every sample of inertial scale length) of the singularity spectrum including possible violent rare events corresponding to negative Hölder exponents. This analysis is currently in progress. It is likely to provide fundamental information about the true role played by the vortex filaments[27, 28] in the intermittency phenomenon of the fine structures of fully developed turbulent flows.

This work was supported by the Direction des Etudes et Techniques under contrat N°92/097.

Bibliography

[1] A.S. Monin and A.M. Yaglom, *Statistical Fluid Mechanics*, vol II (MIT Press, 1975).

[2] U. Frisch and S.A. Orszag, "Turbulence : Challenges for Theory and Experiments", Physics Today (1990), p.24.

[3] A.N. Kolmogorov, C.R. Acad. Sci. USSR **30**, 301 (1941).

[4] B.B. Mandelbrot, J. Fluid Mech. **62**, 331 (1974); *The Fractal Geometry of Nature* (Freeman, San Francisco, 1982).

[5] (a) T.C. Halsey, M.H. Jensen, L.P. Kadanoff, I. Procaccia and B.I. Shraiman, Phys. Rev. A **33**, 1141 (1986); (b) P. Collet, J. Lebowitz and A. Porzio, J. Stat. Phys. **47**, 609 (1987).

[6] (a) R.R. Prasad, C. Meneveau and K.R. Sreenivasan, Phys. Rev. Lett. **61**, 74 (1989); (b) P. Miller and P. Dimotakis, Phys. Fluids A**3**, 168 (1991).

[7] C. Meneveau and K.R. Sreenivasan, J. Fluid Mech. **224**, 429 (1991).

[8] (a) F. Anselmet, Y. Gagne, E.J. Hopfinger and R.A. Antonia, J. Fluid. Mech. **140**, 63 (1984); (b) Y. Gagne, Thesis, University of Grenoble (1987); (c) B. Castaing, Y. Gagne and E.J. Hopfinger, Physica D **46**, 177 (1990).

[9] G. Parisi and U. Frisch, in *Turbulence and Predictability in Geophysical Fluid Dynamics and Climate Dynamics*, edited by M. Ghil, R. Benzi and G. Parisi (North-Holland, Amsterdam, 1985), p. 84.

[10] A.M. Obukhov, J. Fluid Mech. **13**, 77 (1962)

[11] A.N. Kolmogorov, J. Fluid Mech. **13**, 82 (1962)

[12] E. Novikov and R.W. Stewart, Izv. Akad. Nauk. SSSR, Geofiz. **3**, 408 (1964)

[13] U. Frisch, P.L. Sulem and M. Nelkin, J. Fluid. Mech. **87**, 719 (1978).

[14] (a) I. Hosokawa and K. Yamamoto, Phys. Fluid A4, 457 (1991); (b) S.T. Thoroddsen and C.W. Van Atta, Phys. Fluid **A4**, 2592 (1992); (c) S. Chen, G.D. Doolen, R.H. Kraichnan and Z. She, Phys. Fluid A5, 458 (1992).

[15] (a) G. Stolovitsky, P. Kailasnath and K.R. Sreenivasan, Phys. Rev. Lett. **69**, 1178 (1992); (b) J. O'Neil and C. Meneveau, preprint (1992); (c) Y. Gagne, M. Marchand and B. Castaing, J. Phys. II France **4**, 1 (1994).

[16] E. Aurell, U. Frisch, J. Lutsko and M. Vergassola, J. Fluid Mech. **238**, 467 (1992)

[17] A. Bershadskii and A. Tsinober, Phys. Rev. E48, 282 (1993).

[18] J.F. Muzy, E. Bacry and A. Arneodo, Phys. Rev. E47, 875 (1993).

[19] (a) J.F. Muzy, E. Bacry and A. Arneodo, Phys. Rev. Lett. **67**, 3515 (1991); (b) A. Arneodo, E. Bacry and J.F. Muzy, *Wavelets and Turbulence* (Springer, Berlin, 1992) to appear; (c) E. Bacry, J.F. Muzy and A. Arneodo, J. Stat. Phys. **70**, 635 (1993).

[20] (a) *Wavelets*, edited by J.M. Combes, A. Grossmann and P. Tchamitchian, (Springer, Berlin, 1989); (b) *Les Ondelettes en 1989*, edited by P.G. Lemarié (Springer, Berlin, 1990); (c) *Wavelets and Their Applications*, edited by Y. Meyer (Springer, Berlin, 1992); (d) *Progress in Wavelet Analysis and Applications*, edited by Y. Meyer and S. Roques (Editions Frontieres, Gif/Yvettte, 1993).

[21] (a) A. Arneodo, G. Grasseau and M. Holschneider, Phys. Rev. Lett. **61**, 2281 (1988); in ref.[20]a, p. 182; (b) A. Arneodo, F. Argoul and G. Grasseau, in *Nonlinear Dynamics*, edited by G. Turchetti (World Scientific, Singapore, 1988) p. 130; (c) A. Arneodo, F. Argoul, E. Bacry, J. Elezgaray, E. Freysz, G. Grasseau, J.F. Muzy and B. Pouligny, in ref. [20]c, p. 286.

[22] (a) S. Jaffard, C. R. Acad. Sci. Paris **308**, 79 (1989). (b) M. Holschneider and P. Tchamitchian, in ref [20]b, p. 102.

[23] (a) F. Argoul, A. Arneodo, G. Grasseau, Y. Gagne, E.J. Hopfinger and U. Frisch, Nature **338**, 51 (1989); (b) E. Bacry, A. Arneodo, U. Frisch, Y. Gagne and E.J. Hopfinger, in *Turbulence and Coherent Structures*, edited by M. Lesieur and O. Metais (Kluwer, Dordrecht, 1991), p. 203.

[24] (a) M. Vergassola and U. Frisch, Physica D **54**, 58 (1991); (b) M. Vergassola, R.Benzi, L.Biferale and D.Pisarenko, J. Phys. A **26**, 6093 (1993).

[25] S. Mallat and W.L. Hwang, IEEE Trans. on Inf. Th. **38**, 617 (1992).

[26] M. Farge, Annu. Rev. Fluid. Mech. **24**, 395 (1992).

[27] S. Douady, Y. Couder and M. Brachet, Phys. Rev. Lett. **67**, 983 (1991).

[28] (a) E. Siggia, J. Fluid Mech **107**, 375 (1981); (b) M.E. Brachet, D.I. Meiron, S.A. Orzag, B.G. Nickel, R.H. Morf and U. Frisch, J. Fluid. Mech. **130**, 411 (1983); (c) Z.S. She, E. Jackson and S.A Orszag, Nature **344**, 6263 (1990) ; **344**, 226 (1990); (d) A. Vincent and M. Meneguzzi, J. Fluid Mech. **255**, 1 (1991); (e) M.E. Brachet, C. R. Acad. Sci. Paris **311**, 775 (1990) ; Fluid Dyn., **8**, 1 (1991).

[29] B.B. Mandelbrot, Pure Appl. Geophys. **131**, 5 (1989).

Index

MIX
Papier aus verantwortungsvollen Quellen
Paper from responsible sources
FSC® C105338

If you have any concerns about our products,
you can contact us on
ProductSafety@springernature.com

In case Publisher is established outside the EU,
the EU authorized representative is:
Springer Nature Customer Service Center GmbH
Europaplatz 3, 69115 Heidelberg, Germany

Printed by Libri Plureos GmbH
in Hamburg, Germany